TJ 1280 .S27 1992

Salmon, Stuart C.

Modern grinding process
 technology

DATE DUE

DEMCO 38-297

NEW ENGLAND INSTITUTE
OF TECHNOLOGY
LEARNING RESOURCES CENTER

Modern Grinding
Process Technology

Other Books of Interest

BAUMEISTER & MARKS • *Marks' Standard Handbook for Mechanical Engineers*
BHUSHAN & GUPTA • *Handbook of Tribology*
BRADY & CLAUSER • *Materials Handbook*
BRALLA • *Handbook of Product Design for Manufacturing*
BRUNNER • *Handbook of Incineration Systems*
CORBITT • *Standard Handbook of Environmental Engineering*
EHRICH • *Handbook of Rotordynamics*
ELLIOT • *Standard Handbook of Powerplant Engineering*
FREEMAN • *Standard Handbook of Hazardous Waste Treatment and Disposal*
GANIC & HICKS • *The McGraw-Hill Handbook of Essential Engineering Information and Data*
GIECK • *Engineering Formulas*
GRIMM & ROSALER • *Handbook of HVAC Design*
HARRIS • *Handbook of Noise Control*
HARRIS & CREDE • *Shock and Vibration Handbook*
HICKS • *Standard Handbook of Engineering Calculations*
HODSON • *Maynard's Industrial Engineering Handbook*
JONES • *Diesel Plant Operations Handbook*
JURAN & GRYNA • *Juran's Quality Control Handbook*
KURTZ • *Handbook of Applied Mathematics for Engineers and Scientists*
KARASSIK, ET AL. • *Pump Handbook*
PARMLEY • *Standard Handbook of Fastening and Joining*
ROHSENOW, ET AL. • *Handbook of Heat Transfer Fundamentals*
ROHSENOW, ET AL. • *Handbook of Heat Transfer Applications*
ROSALER & RICE • *Standard Handbook of Plant Engineering*
ROTHBART • *Mechanical Design and Systems Handbook*
SCHWARTZ • *Composite Materials Handbook*
SCHWARTZ • *Handbook of Structural Ceramics*
SHIGLEY & MISCHKE • *Standard Handbook of Machine Design*
TOWNSEND • *Dudley's Gear Handbook*
TUMA • *Handbook of Numerical Calculations in Engineering*
TUMA • *Engineering Mathematics Handbook*
WADSWORTH • *Handbook of Statistical Methods for Engineers and Scientists*
YOUNG • *Roark's Formulas for Stress and Strain*

Modern Grinding Process Technology

Dr. Stuart C. Salmon

McGraw-Hill, Inc.
New York   St. Louis   San Francisco   Auckland   Bogotá
Caracas   Lisbon   London   Madrid   Mexico   Milan
Montreal   New Delhi   Paris   San Juan   São Paulo
Singapore   Sydney   Tokyo   Toronto

Library of Congress Cataloging-in-Publication Data

Salmon, Stuart C.
   Modern grinding process technology / Stuart C. Salmon.
      p.   cm.
   Includes index.
   ISBN 0-07-054500-6
   1. Grinding and polishing. I. Title.
TJ1280.S27 1992
621.9'—dc20   91-45421
                                                                CIP

Copyright © 1992 by McGraw-Hill, Inc. All rights reserved. Printed in the United States of America. Except as permitted under the United States Copyright Act of 1976, no part of this publication may be reproduced or distributed in any form or by any means, or stored in a data base or retrieval system, without the prior written permission of the publisher.

1 2 3 4 5 6 7 8 9 0   DOC/DOC   9 8 7 6 5 4 3 2

ISBN 0-07-054500-6

The sponsoring editor for this book was Robert Hauserman, and the production supervisor was Suzanne W. Babeuf. It was set in Century Schoolbook by North Market Street Graphics.

Printed and bound by R. R. Donnelley & Sons Company.

Amborite ABN is a trademark of DeBeers, Ascot, England.
Borazon is a trademark of GE Superabrasives, Worthington, OH, USA.
Cubitron is a trademark of 3M Company, St. Paul, MN, USA.
Diamesh is a trademark of Ultimate Abrasive Systems, Atlanta, GA, USA.
Granitan is a trademark of Studer AG, Thun, Switzerland.
Grind-O-Sonic is a trademark of J.W. Lemmens Inc., Anaheim, CA, USA.
Man-Made is a trademark of GE Superabrasives, Worthington, OH, USA.
Norbide is a trademark of Norton Company, Worcester, MA, USA.
Roll-2-Dress is a trademark of Ernst Winter and Sohn, Hamburg, Germany.
Swing-Step is a trademark of KW Antriebstechnik GmbH, Mainaschaff, Germany.

Information contained in this work has been obtained by McGraw-Hill, Inc. from sources believed to be reliable. However, neither McGraw-Hill nor its authors guarantees the accuracy or completeness of any information published herein and neither McGraw-Hill nor its authors shall be responsible for any errors, omissions, or damages arising out of use of this information. This work is published with the understanding that McGraw-Hill and its authors are supplying information but are not attempting to render engineering or other professional services. If such services are required, the assistance of an appropriate professional should be sought.

I dedicate this book to my father, Arthur Henry Salmon, who died in England on May 1, 1991. Along with my mother he was the driving force behind my entire life. He challenged me. There was always a better way. Nothing was ever good enough. It seemed, at times, that there was little that could please him; it was then he would encourage me. He always wanted the best. Dad, I tried my hardest and will keep on trying. I love you and I miss you.

Contents

List of Figures   xi
Preface   xv
Acknowledgments   xvii

Chapter 1. An Introduction to Abrasives   1

1.1   Abrasives in Our Everyday Lives   2
1.2   The History of Abrasives   2
1.3   The World Market in Abrasives   6

Chapter 2. The Manufacture and Properties of Abrasives   9

2.1   Introduction   9
2.2   Silicon Carbide   9
2.3   Aluminum Oxide (Fused)   12
2.4   Aluminum Oxide (Ceramic)   14
2.5   Cubic Boron Nitride (CBN)   15
2.6   Diamond   19
2.7   The Properties of Abrasives   21
2.8   Concentration   23
2.9   Vitrified-Bonded Grinding Wheels   26
2.10   Resin-Bonded Grinding Wheels   30
2.11   Rubber-Bonded Grinding Wheels   31
2.12   Metal-Bonded Grinding Wheels   31
2.13   Abrasive Belts—Coated Abrasives   34
2.14   Backing Materials for Coated Abrasives   36
2.15   Adhesives for Coated Abrasives   36
2.16   The Manufacture of Coated Abrasives   37
2.17   The Storage of Coated Abrasives   39

Chapter 3. Abrasive Preparation   41

3.1   Grinding Wheel Preparation—Mounting   41
3.2   Fitting a Coated Abrasive Belt   50
3.3   Grinding Wheel Conditioning   50

3.4	Single Point Dressing	55
3.5	Crush Dressing	64
3.6	Diamond Roll Dressing	67
3.7	Continuous Dressing	85
3.8	EDM Dressing Metal Bond Systems	85

Chapter 4. Fundamentals of Grinding — 89

| 4.1 | A Micro-milling Analogy | 89 |
| 4.2 | Energy Used for Grinding | 94 |

Chapter 5. Grinding Machine Tool Design — 103

5.1	Introduction	103
5.2	An Historical Perspective	104
5.3	Creep-feed Grinding Machine Design	106
5.4	High-speed Grinding Machine Tool Design	109
5.5	Vibration in Machine Tools	110
5.6	The Next Generation Grinding Machine	112
5.7	Computer Numerical Control (CNC)	116

Chapter 6. Cutting Fluid Application and Filtration — 121

6.1	The Role of the Cutting Fluid	121
6.2	Filtration of the Cutting Fluid	127
6.3	Types of Cutting Fluid	129

Chapter 7. Cylindrical Grinding Processes — 133

7.1	Outside Diameter (OD) Grinding	133
7.2	Plain Cylindrical Grinding	135
7.3	Plunge Cylindrical Grinding	142
7.4	Angle Approach Grinding	149
7.5	Internal Diameter (ID) Grinding	153
7.6	Cam Grinding	153
7.7	In-process Gaging	157

Chapter 8. Flat Surface Grinding Processes — 161

8.1	Reciprocating Grinding	161
8.2	Form Surface Grinding Processes	172
8.3	Creep-feed Grinding	175
8.4	Speed-feed Grinding	181

Chapter 9. Special Grinding Processes — 185

9.1	Centerless Grinding	185
9.2	Tool and Cutter Grinding	188
9.3	Electrolytic Grinding	190

9.4	Honing and Super-finishing	192
9.5	Snagging and Cut-off	194
9.6	Double-disk Grinding	195
9.7	Lapping and Free-abrasive Machining	199

Chapter 10. Coated Abrasive Processes   201

Chapter 11. Surface Finish and Integrity Measurement   207

  Glossary   215
  Index   223

List of Figures

CHAPTER 1

1.1	Market value of ceramics.	7

CHAPTER 2

2.1	Standard wheel designations.	10
2.2	A piece of silicon carbide broken from the cooled charge.	12
2.3	A piece of aluminum oxide broken from the cooled charge.	13
2.4	Properties of aluminum oxide and silicon carbide.	14
2.5	Seeded-Gel (SG) aluminum oxide and fused aluminum oxide.	16
2.6	Nickel-coated CBN 560 with CBN 550, an uncoated abrasive.	18
2.7	Grain wear.	20
2.8	Grain size.	24
2.9	Grain concentration.	25
2.10	Grind-O-Sonic.	27
2.11	Grinding wheel structure.	29
2.12	A rim section of a vitrified CBN grinding wheel.	31
2.13	Type S/1 diamesh which has a single abrasive particle per mesh opening and type M/1 diamesh with two abrasive particles per mesh opening.	33
2.14	Methods for storing grinding wheels.	35
2.15	Cross section of a coated abrasive product.	36
2.16	Lap joint in a 100 mm (4 in) wide abrasive belt.	37
2.17	Coated abrasive designation system.	38

CHAPTER 3

3.1	Grinding wheel assembly.	42

3.2	Bolt tightening sequence for six and eight bolt flanges.	43
3.3	Balancing a grinding wheel on knife edges.	44
3.4	Manually balancing a grinding wheel.	46–47
3.5	Fluid injection balancing.	48
3.6	Electro-mechanical dynamic balancing.	49
3.7	Truing a metal-bonded wheel.	51
3.8	A brake-controlled dresser used to true a resin-bonded superabrasive wheel.	53
3.9	Dressing a resin-bonded wheel by sticking.	55
3.10	Roll-2 dressing system.	56
3.11	Pantograph and CNC dressing.	57
3.12	Computer-controlled dressing.	58–59
3.13	Single-point diamond dressing.	60
3.14	Fliese tool dressing.	61
3.15	Principles of single point dressing by overlays.	63
3.16	Diamond wafer roll dressing.	65
3.17	Comparison of forces when crush dressing and diamond roll dressing.	66
3.18	Diamond roll dressing.	69
3.19	Diamond roll dressing unit.	70
3.20	Types and construction of diamond rollers.	72
3.21	Accuracy of diamond rolls.	73
3.22	Diamond roll mounting.	75
3.23	Cross section of a diamond roll dressing unit.	77
3.24	Swing-step dresser.	78
3.25	The effect of dressing speed ratio on specific energy.	79
3.26	Overhead dressing system.	80
3.27	Critical dwell period.	81
3.28	Diamond roll removal.	82
3.29	Diamond roll wear.	83
3.30	Optimized dressing system for superabrasive wheels.	84
3.31	Continuous diamond roll dressing.	86
3.32	Electro-discharge machine (EDM) dressing.	87

CHAPTER 4

4.1	Geometric diagram of the micro-milling analogy.	90
4.2	Three modes of grinding energy.	96
4.3	Comparison of grinding forces—ceramic vs. metal.	100

CHAPTER 5

5.1	Reciprocating surface grinding machine.	105
5.2	Single wheel and dual spindle surface grinding machines.	108

5.3	Casting and weldment cross sections.	111
5.4	Surface and cylindrical NGGM concept.	114
5.5	Combination moving column and rotary table machine.	115
5.6	De-coupled, rotary table NGGM.	116
5.7	Coupled NGGM.	117
5.8	Knife grinder.	118

CHAPTER 6

6.1	Sharp abrasive vs. dull abrasive.	122
6.2	Film boiling.	123
6.3	The effect of the cutting fluid temperature on the stock removal rate.	124
6.4	Cutting fluid application method for high-speed cylindrical grinding.	126
6.5	Fixture design necessary to maintain good cutting fluid flow.	128
6.6	Cutting fluid cooling systems.	130

CHAPTER 7

7.1	Basic cylindrical grinding machine configuration.	134
7.2	Rotation direction of the grinding wheel and workpiece in cylindrical grinding.	135
7.3	The swivelling table allows cylindrical grinding of tapers.	136
7.4	The dog drive.	137
7.5	Angle approach grinding.	139
7.6	Taper bore grinding.	140
7.7	Cylindrical grinding.	141
7.8	Distribution of grinding forces across a grinding wheel during cylindrical grinding.	144–145
7.9	Multi-grade grinding wheel.	146
7.10	Plunge grinding.	147
7.11	Steady-rests.	148
7.12	Two types of grinding marks on a ground face.	150–151
7.13	Regenerative chatter.	152
7.14	The quill shaft for ID grinding.	154
7.15	Clamping force.	155
7.16	Cam grinding.	156
7.17	In-process gaging.	158

CHAPTER 8

| 8.1 | Reciprocating grinding. | 162 |
| 8.2 | Pre-loaded roller guideway bearing system. | 164 |

8.3	V-flat guideway bearing system.	166
8.4	Hydrostatic bearings.	168–169
8.5	The vibrational stability of epoxy granite and cast iron.	170
8.6	Vibrational instability of a grinding wheel.	171
8.7	Vertical spindle grinding system.	173
8.8	Rotary surface grinding system.	174
8.9	Reciprocating plunge form grinding.	176
8.10	Conventional reciprocating grinding and creep-feed grinding.	177
8.11	Twin spindle multi-axis grinding center.	179
8.12	CNC creep-feed grinding.	181
8.13	Speed-feed grinding.	182

CHAPTER 9

9.1	The centerless grinding mechanism.	186
9.2	Centerless grinding a taper to an end stop.	187
9.3	Tool and cutter grinding.	188
9.4	Electrolytic grinding system.	191
9.5	Rotary honing and superfinishing.	193
9.6	Cut-off grinding.	195
9.7	Double disk grinding.	196
9.8	Lapping and free abrasive machining.	197
9.9	Vibratory finishing.	198
9.10	Examples of finishing media materials, size and shape.	199

CHAPTER 10

10.1	Coated abrasive machines and their applications.	202–203
10.2	Form grinding with coated abrasives.	205

CHAPTER 11

11.1	Scanning electron micrographs of CBN grain.	209
11.2	Surface finish measurement.	210–211
11.3	Working surfaces.	212
11.4	Surface finish/manufacturing cost.	213

Preface

This book has been distilled from more than twenty years of both practical experience and academic research in production grinding processes and systems.

My first attempt at writing an up-to-date grinding manual came with Korber AG in Hamburg, Germany. At their request a text book was written mainly to promote their Schaudt cylindrical and Blohm surface grinding machine tool products and technology. The *Abrasive Machining Handbook* was produced by Korber AG and given free to its customers.

Know-how is the key to success. The *Abrasive Machining Handbook* provided insight into the latest grinding processes and gave confidence to an industry which was hanging on to the historical "trial and error" practices from a bygone time.

The Society of Manufacturing Engineers saw Dr. Robert Hahn retire from the lecture circuit and wanted to provide the United States industry with a Modern Grinding Technology Workshop encompassing the most modern grinding techniques. A course book was written for the workshop attendees drawing on the basic fundamentals to take the audience into the next generation of grinding process technology. The book underwent two iterations and forms the backbone of this text.

It is the aim of this book to be a reference guide providing both practical and theoretical understanding so that the reader can be confident in his or her approach to applying modern grinding technology to the manufacturing world of materials today and tomorrow.

I hope that you find the book useful and easy to read. Ultimately you will have to put the information to good use. There are exciting things happening in the world of abrasives: new processes, new tools, and more efficient methods. We can no longer keep on trying to do the same old things better. There is much to try to do and it all begins just a few pages from here.

Acknowledgments

In order to write a book such as this, there has to be a commitment to impart truth and confidence in a noncommercial way in a commercial world. The abrasives industry is entering exciting times. I am most grateful to the following companies for their assistance in this work, supplying photographs, graphics, and comments. Without them this book would be incomplete and without each other we have nothing. Thank you:

3M Company—St. Paul, Minnesota

6th Grinding Wheel Factory—Guizhou Province, China

Abrasive Engineering Society—Butler, Pennsylvania

American Ceramics Society—Westerville, Ohio

Brown & Sharpe—Providence, Rhode Island

Cincinnati-Milacron—Products Division, Cincinnati, Ohio

De Beers—Ascot, England and Diamond Abrasive Corporation, New York, New York

Diamond Winter—Hamburg, Germany and Travelers Rest, South Carolina

ELB—Babenhausen, Germany and Fort Lauderdale, Florida

GE Superabrasives—Worthington, Ohio

Hauni-Blohm—Hamburg, Germany and Richmond, Virginia

Jones and Shipman—Leicester, England and Atlanta, Georgia

K.W. Antriebstechnik GmbH—Mainaschaff, Germany and Charlotte, North Carolina

Maegerle—Uster, Switzerland and Schaumberg, Illinois

Master Chemical Corporation—Perrysburg, Ohio

Norton Company—Worcester, Massachusetts

Rolls-Royce Apprentice Training School—Derby, England

SBS—Portland, Oregon

Schaudt—Stuttgart, Germany and Dayton, Ohio

Society of Manufacturing Engineers—Dearborn, Michigan

Studer—Thun, Switzerland and Brookfield, Connecticut

Tyrolit—Schwaz, Austria and Newport News, Virginia

Ultimate Abrasive Systems—Atlanta, Georgia

Wendt GmbH—Meerbusch, Germany and Wendt Dunnington, Chester Springs, Pennsylvania

Winterthur Grinding Wheel Company—Winterthur, Switzerland and Worcester, Massachusetts

Last and by no means least, I extend a special thank you to my wife Sue, who, throughout the ordeal of compiling this book, writing and rewriting, late nights, early mornings, and cancelled weekends, complained little. In spite of all that stood in our way, you were my strongest source of encouragement to press on. On behalf of all those who benefit from this work, thank you.

Modern Grinding
Process Technology

Chapter 1

An Introduction to Abrasives

Abrasives and the action of abrasion have been in progress since the beginning of time. Since then, the earth has taken on an ever-changing form due to the erosion of rocks and land mass. Some erosion is caused by the wind and some by the ocean. We will, however, discount human attempts to rearrange the planet. The wind may carry particles of dust and sand, and the ocean and water may carry rocks, pebbles, and boulders. The air and the water are the transporting media moving the harder particles across and against the softer land masses. These are abrasives at work, creating wonderful and amazing natural rock formations like the Grand Canyon, Natural Arch Bridge, and Watkins Glen. We might even consider the craters on the moon, caused by an abrasive action from the bombardment of meteoroids from outer space. In some cases the gentle erosion by the wind has created very picturesque features on our land; in other cases, the natural abrasive action has brought disaster in the form of erosion at the coastline, washing away entire communities at the water's edge. Such is the power of the abrasive action in nature and, as we shall learn through the course of this book, similarly so in industry.

Stone Age people were the first "Abrasive Engineers." They formed weapons and tools, shaping them by abrasion. From the natural stones and rocks found around them, prehistoric humans saw that it was not the hard stones that ground the best, but the softer, more fragile sandstones. This is a phenomenon that will develop in our understanding as we progress through the course. Abrasives, it seems, are almost an inherent part of our being. We use them in our everyday lives without even giving them a second thought.

1.1 Abrasives in Our Everyday Lives

Every morning when we brush our teeth we use abrasives. Most toothpastes contain a very mild abrasive like hydrated silica, which helps clean our teeth, abrading away deposited plaque. The shower stall, bathtub, and washbasin may well have been cleaned using household products containing silica or calcium carbonate, a milder abrasive. The metal utensils with which we eat have been finished using abrasives. The bicycle, motorcycle, or automobile we use to transport ourselves and the electronics and telecommunication equipment we use to do business would be very difficult to make, but for the proper use of abrasives. The foods we eat, flour and chocolate for example, all require the direct use of abrasives in their manufacture. The furniture in our homes was finished using abrasive processes and abrasive media. Surprising as it may seem, our lives depend almost totally on abrasives. Therefore, it would be sensible to understand their uses, as well as how to use them safely, economically, and efficiently.

All of us, at one time or another, have sharpened a pencil or a crayon on a rough stone, concrete, or brick surface. Inherently, we did not try to shape the brick or the stone with the pencil. The most basic principles of abrasives seem to be born in us. The abrasive grain must be aggressive and sharp. Moreover, the abrasive grain has to be harder than the material we wish to shape. Materials scientists are developing materials with hardness and wear resistance which pose a significant challenge to the manufacturing engineer. As materials science develops these new materials, so new abrasive products and processes have to be researched and developed. A whole new world of "Superabrasives" now exists, made from ultra-hard materials, some artificial and some naturally occurring, like diamonds, which are used to machine and fabricate the next generation of materials.

1.2 The History of Abrasives

Let us take a moment to look back and imagine the work of the first engineers. In particular, we will focus on the likes of James Watt and George Stevenson. They pioneered the era of the steam engine, the railroad engine, and of course the whole railroad and transportation system. To build a successful steam engine, precision engineering was a prerequisite. The valve gear and the engine piston bores and pistons had to have close sliding fits. Tolerancing was in its infancy and standards of limits and fits and interchangeability were a thing for the future. Each engine's piston and bore was machined individually and painstakingly formed to a shape which gave the required engine life and performance. The industrial revolution caused the engineering profession to rethink and adjust its mission. No longer was striving for perfection the single purpose in an engineer's life. Of equal consideration and importance was the production of the commodity and the interchangeability of parts manufactured efficiently and economically. With Henry Ford came the motor car and mass production. Interchangeability and exacting engineering standards demand a tolerancing system of clearances and limits and fits of assembled components. Creating accurate fits requires very precise machining. Milling, broaching, and turning are generally not accurate enough for today's demands in precision. Grinding is an essential process not only for precise sizing and form accuracy, but also for producing the necessary working surface finish. From those early times of Watt, Stevenson, and Ford it was recognized that the demand for consistency and better control of size and surface finish were essential to the success and continuing improvements in engineering design and production. Abrasive engineering delivers that result. Surprisingly, abrasives have been under our noses since the beginning of time.

Over the course of this book we will concentrate on the use of abrasives and develop a practical understanding of their origins, their manufacture, their preparation, and the processes used in modern-day industry.

Abrasive machining has become the generic name given to the process of removing materials, both metals and nonmetals, by means of making very small chips or particulate produced by the cutting edges of abrasive particles. Abrasive papers (sandpaper, emery paper, etc.) first appeared on the streets of Paris, France in the late eighteenth century. The first known article describing the method of making coated abrasives appeared in 1808; calcinated and ground pumice were mixed together with a varnish and spread onto paper with a brush. Emery cloth was invented in England in 1831 (sand, powdered glass, or emery was mixed with glue and spread onto a cotton cloth). An American patent in 1835 covered a method for making coated abrasives by allowing steam to act on the reverse, uncoated, side of the paper to prevent curling. Sand was sprinkled from a sieve onto glue-coated paper that was carried by an endless belt. Prior to and during this time, naturally occurring elements such as sandstone and corundum were cut from solid rock to form the rounded shape of a grinding wheel to be used with newly developed mechanisms and machines.

The abrasive particle is a very hard material. It can only cut or abrade other materials which are softer than itself. By virtue of its shape and its hardness, the abrasive particle is able to penetrate the surface of the material being machined. Since ancient times, natural abrasives were plentiful and were used extensively to sharpen tools and weapons as we previously described. The sandstone wheels were not homogeneous and wore unevenly. Remember, the stone itself was soft, but it was the features of abrasive particle, its shape and its hardness, which allowed abrasion to take place. The inconsistencies in the natural stone, however, meant that the essentially manual machining process was most unpredictable and yielded less than satisfactory results. The need for consistency in an abrasive tool led to the manufacture of artificial abrasives which are more prevalent today. Natural abrasives still in use today are emery, diamond, and quartz sand. The emery and quartz are typically bonded to paper or cloth and form the familiar sandpaper and emery paper used in woodworking. Crocus rouge is still used in very fine polishing compounds, as well as diamond in the form of very fine dust. A common application for such polishing compounds is in metallography for the preparation of the flat surface of the specimen mount.

As the industrial revolution progressed, the need for stronger and more consistent grinding wheels grew. No longer were the unpredictable natural forces holding a sandstone together reliable, particularly as grinding wheel speeds became faster. As a first step, grinding wheels were manufactured by crushing the natural stones into grains and reforming them into grinding wheels using clays and resins to bond the abrasive grain together again. The resin wheels were baked to cure them and the clays were fired in a kiln just like porcelain and china is made today. Because the wheels were made from very small particles, they could be molded into a whole variety of shapes. The use of abrasives was forming into three distinct groups—bonded, coated, and free abrasives. Bonded abrasives are held together by a bonding agent to form a grinding wheel or stone. Coated abrasives are adhered to papers and cloths in the form of grinding belts. Free abrasives are suspended in a fluid medium for polishing or tumbling.

It was not until the nineteenth century that synthetic abrasives began to replace the natural abrasives of sandstone, crocus rouge, emery, corundum, and diamond. The natural abrasives which occurred due to the forces of nature contained many impurities and varied in quality. Synthetic abrasives, however, are pure, consistent, and can be carefully controlled. The most common artificial abrasives available today, in order of their popularity, are aluminum oxide, silicon carbide, cubic boron nitride (CBN), and synthetic diamond. Aluminum oxide and silicon carbide combined are used in over 90 percent of today's abrasive machining applications. Cubic boron nitride and synthetic diamond, termed "Superabrasives," form only a small part of the overall abrasive usage due to their cost and the very special considerations necessary for their successful and economical application. Nonetheless, the new "Superabrasives" are enjoying wider acceptance and are growing in their application. CBN is particularly popular in the Pacific Rim nations. In fact, it is estimated that in the 1980s over 60 percent of all CBN produced was used in Japan.

Today, the abrasive grain is virtually all artificial and can be manufactured in a variety of different, carefully controlled shapes and sizes. The ability to make abrasive grain consistently to high levels of performance and accurate size provides the abrasive machining process with a wide range of success in a world of machining operations. Machining with abrasives may be carried out with extremely high precision. There are specialist groups of engineers and scientists dedicated to nano-precision (10^{-9}). Nano-precision signifies dimensional tolerancing within a nanometer (0.00000004 in). Nanometer precision furthers the science of achieving such levels of dimensional accuracy and also provides exceptionally high degrees of surface finish. Generally, for high precision work, the stock removal rate in abrasive machining is small due to the size of the abrasive particle being very small. Conversely, abrasive machining may also be used to remove large quantities of material in a very short time, generally to the detriment of surface finish and high precision. Yet, as we shall see, there are some abrasive processes which are exceptions to the rule. The world of abrasive machining covers both ends of the spectrum. Indeed, abrasive machining can achieve the highest qualities of surface finish better than any other machining process. Almost any shape or size of workpiece may be finished by abrasive machining. Surface grinders produce flat, angular, or contoured forms on a flat or contoured surface. Cylindrical grinders produce both internal and external cylindrical shapes, tapers, thread forms, and cam profiles. Recent developments in abrasives and abrasive processes have proven that the complete machining of a workpiece from a rough casting or forging can be achieved by grinding processes only. In most cases grinding achieves better results, both economically and in terms of surface integrity, without the need to premachine the workpiece by conventional milling, broaching, or turning processes. Abrasive machining has a most exciting future. Today it is the only process available which successfully and economically machines the most difficult superalloys. With the ever increasing number of ceramic materials in the marketplace, abrasive machining is assured a place in machining technology. There are few alternatives. One cannot technically mill, broach, or turn these very hard and brittle ceramics with any economical advantage. Grinding is both technically and economically the number one choice when one has to consider machining ceramics. As we look to the future, there is little doubt that abrasive machining is a most versatile process with a very prosperous future.

The abrasive particle is a cutting tool and a most important part of the abrasive machining system. Each abrasive type has different physical properties, as well as both practical and economic considerations. The objective of today's manufacturing world is to achieve the lowest piece part cost for the desired quality and quantity of those parts. Abrasive cost, tooling, and equipment cost are critical in this scenario. The cost of labor, however, is fast becoming less significant in the face of automation and the computer control of the process.

It is therefore important that we take an in-depth look at the major abrasives in use with today's industry so that we might develop an understanding and an approach to selecting the most suitable abrasive media and abrasive processes to achieve the lowest piece part cost for the appropriate level of part quality.

1.3 The World Market in Abrasives

Before analyzing the grain, its manufacture and its uses, we should appreciate the size of the world market in abrasives. Going into the 1990s the American sales of abrasive products was over $1 billion. The breakdown of this figure is roughly $650 million in coated abrasive products, $400 million in conventional wheels, and $120 million in superabrasives. The western world combined adds another $1.75 billion to the total annual abrasive products sold in the free world. It is an interesting statistic that about 30 percent of all machine tools are abrasives oriented. If the sales of abrasive machine tools and abrasive consumable products are combined for the United States, the total market is in excess of $2.5 billion annually. World trends show us that:

>   *Europe and Scandinavia*—They parallel the United States in dollar volume of abrasive sales. Along with Japan, Europe has pioneered much of the CBN application development, as well as leading the way in new creep-feed grinding technology and automated grinding systems. Most major abrasive companies and suppliers in Europe are firmly established in the United States. The reverse is also true for U.S. suppliers in Europe.
>
>   *Canada*—Canada is an appreciable user and supplier of abrasive products with particular strength in coated abrasives and wide belts for forestry applications.
>
>   *Peoples Republic of China*—China has a complete and self-supporting abrasives industry producing abrasives of all types, as well as a strong machine tool industry. China exports raw abrasive materials—aluminum oxide, silicon carbide, zirconia, CBN, and synthetic diamonds—to the rest of the world.
>
>   *The Pacific Rim*—Japan, Korea, and Taiwan have a potential market annually in excess of $350 million in abrasive products in the 1990s. Japan has produced some of the finest machine tools, specifically

designed with superabrasives in mind. The Pacific Rim nations are also among the world leaders in ceramic technologies, both in processing and their manufacture into consumer products. It is here, on the Pacific Rim, where we will witness the upsurge in the application of advanced abrasive machining technology into the 21st Century.

Australia, India, and the Far East—Much like Canada, coated abrasives for the forestry and woodworking industries take the lions' share of the estimated $425 million abrasives market. $175 million of that market is in bonded abrasives for precision grinding in areas like Malaysia, Indonesia, and the Philippines, where high technology manufacturing is increasing rapidly. India has an established precision grinding industry and conducts significant research and development in abrasive technology.

South America—Brazil has a completely independent and self-sufficient abrasives industry, and exports a large volume of abrasive

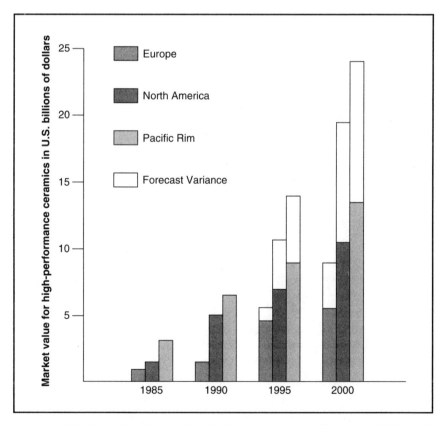

Figure 1.1 Market value of ceramics. (*Reference courtesy of Industrial Diamond Review—January 1991.*)

products to the rest of the world. Argentina has a growth rate in abrasives of 15 percent annually since 1983. Due to the very low cost of labor in the South American countries, much of the mundane machining tasks are being relocated there from North America, boosting the abrasives market in that region.

Eastern Bloc—The Eastern Bloc countries conduct much basic research and development in their universities and research institutes. Much of the equipment and machine tools used in the Eastern Bloc countries comes from Europe, as few highly sophisticated machine tools are built in these Eastern Bloc countries. This situation is changing, but very slowly. The pace of technological change is not keeping up with the sweeping political reforms of the late 1980s.

The ceramics industry is seen as the next frontier in abrasive machining with a market potential of phenomenal proportions (see Fig. 1.1).

Chapter 2

The Manufacture and Properties of Abrasives

2.1 Introduction

In 1944 the Grinding Wheel Manufacturers Association adopted a standard marking system to identify the characteristics of a grinding wheel. In 1958 the American Standards Association adopted a version of the standard. Grinding wheels which contain superabrasives, either CBN or diamond, and particularly those manufactured in Europe, tend to have a marking system which adheres more to the European *Federation Europeenne des Fabricants de Produits Abrasifs* (FEPA) standard (see Fig. 2.1). Throughout this chapter, reference will be made to the standard wheel designation symbols in line with the properties of the various abrasives and particularly the way in which a grinding wheel is manufactured.

There are four main types of synthetic abrasives: aluminum oxide, silicon carbide, Cubic Boron Nitride (CBN), and synthetic diamond. We will examine the procedures for manufacturing these abrasives and the properties of each of them.

2.2 Silicon Carbide

- Chemical formula—SiC.
- Standard wheel designation symbol C.

The first abrasive to be manufactured was silicon carbide (SiC). In 1891 Dr. Edward G. Acheson, working on the assumption that diamond is indeed a form of carbon, mixed together powdered coke and

Standard US grinding wheel designation.

Manufacturer's Prefix and abrasive type	Grain size	Hardness	Porosity	Bond type
38A	**60**	**K**	**2**	**V**

A - Aluminum Oxide	Size in US Mesh.	Alphabetical	Dense 0-3	V - Vitrified
C - Silicon Carbide	Coarse 16-36	A - Softest	Medium 4-6	B - Resin
B - Cubic Boron Nitride	Medium 46-80	Z - Hardest	Open 7-9	R - Rubber
D - Diamond	Fine 90-220		Highly Porous 10+	M - Metal

European/US standard for superabrasive wheels.

Due to the influx of superabrasive wheels from Europe and Japan, the grading system has become mixed and will need some "detective work" to sort out. An example is as follows:

D126   R   75   B

FEPA standard Diamond 120/140 US mesh.	R Hardness	75 Concentration	B - Resin bond

FEPA equivalents:

FEPA	US Mesh
601	30-35
501	35-40
426	40-45
356	45-50
301	50-60
251	60-70
213	70-80
181	80-100
151	100-120
126	120-140
107	140-170
91	170-200
76	200-230
64	230-270
54	270-325
46	325-400

Figure 2.1   Standard wheel designations.

sand and fused the mixture in a crude electric arc furnace. He produced silicon carbide. Acheson thought that he had discovered a substance composed of carbon and corundum (a popular natural abrasive of the time). He named his substance Carborundum and later founded the Carborundum Company.

Silicon carbide is manufactured on an industrial scale today in a resistance arc furnace. The furnace is a refractory enclosure, typically 3 m (10 ft) high, 3 m (10 ft) wide and up to 12 m (40 ft) long with a carbon graphite electrode entering the furnace from either end. The mixture is made of 60 percent silica sand and 40 percent coke, along with a small amount of sawdust, to increase the porosity of the mix. Salt is also added as a catalyst to assist in the purification of the silicon carbide. The furnace is half filled with the mixture and a central core of granular carbon is laid down to connect the two carbon electrodes at either end of the furnace. The furnace is then completely filled. Some furnaces might contain as much as 90,000 kg (200,000 lb) of mix which could yield up to 11,000 kg (25,000 lb) of silicon carbide. A voltage of approximately 300 V is applied to the electrodes and the current allowed to pass for up to 36 hours, over which time the voltage drops to 200 V. During this time the furnace reaches temperatures over 2200°C (4000°F). The furnace is cooled after a soaking time at temperature and the side walls of the furnace are dismantled to expose the charge. The charge is then broken and crushed. The center core of graphite is usually saved to be reused. The center of the charge offers the most pure silicon carbide which is green in color, while the less pure abrasive is black. The need for large amounts of electrical power to energize the furnaces tends to make geographical location important to the producers. China and Scandinavia are large producers of silicon carbide due to the abundance of hydroelectric power as a natural resource. Indeed, the original Carborundum Company settled in Niagara Falls, New York, one of the United States' great hydroelectric power resources.

Silicon carbide (see Fig. 2.2) is especially suited to the machining of very hard materials, in particular tungsten carbide, cast iron, chilled iron, marble, and some ceramics, as well as the more reactive yet softer materials like titanium, aluminum, copper, and brass. Silicon carbide is very hard and has an aggressive shape which assists in penetrating the surface of softer materials like low tensile strength steels, aluminum and copper alloys, plastics, rubbers, and soft wood. The slightly impure black silicon carbide is well suited to very rough grinding and snagging operations, whereas the pure green silicon carbide is more friable and lends itself to cooler cutting and more precise grinding operations in very hard and heat sensitive materials.

Figure 2.2 A piece of silicon carbide broken from the cooled charge.

2.3 Aluminum Oxide (Fused)

- Chemical formula—Al_2O_3.
- Standard wheel designation symbol A.

Around the same time as Acheson, in 1899, Charles B. Jacobs was performing similar experiments to produce pure and more uniform aluminum oxide in France.

The industrial manufacture of aluminum oxide requires a fusion process similar to that for silicon carbide. Bauxite, a naturally occurring clay-like substance, mined from large open-cast mines, is rich in hydrated aluminum oxide. Before processing, the hydrated aluminum oxide has to be calcinated to drive off any moisture. The bauxite is then mixed with ground coke and iron borings. The base of an open cylindrical pot furnace is covered with carbon bricks. The sides are left uncovered, however, and when in operation they are cooled by water jets. Once the furnace is half filled with the above mixture, two or three vertical electrodes are lowered into the furnace. A starter charge of metallurgical coke is placed between the electrodes and the mixture. The electric current is applied and the coke quickly glows to incandescence. The fusion process has begun. The intense heat, in the order of 2000°C (3700°F), melts the bauxite and reduces the impurities which settle to the bottom of the charge. As the fusion process continues, more mixture is added until the furnace is full. The electrodes are raised and lowered to maintain the correct temperature automatically. This fusion process may last from 16 to 36 hours. Once complete, the furnace is left to cool for several days. The ingot is then emptied of its charge and the outer impure layer is stripped off. The center core of 99.9 percent pure aluminum oxide is broken up and crushed. Once crushed, the abrasive is washed and then screened for size. Pure aluminum oxide is white (see Fig. 2.3). Regular aluminum oxide is a special term given to impure aluminum oxide which is brown in color and typically 95 percent pure. The hardness and brittleness of the abrasive increases with purity. Aluminum oxide is the most commonly used abrasive and is used to machine a vast array of materials from the most difficult machine superalloys to high alloy steels and mild steels.

There are a number of techniques used to modify the properties of the aluminum oxide grain to adjust its friability, hardness, and bonding. The silicon treatment is one which improves the grain's adhesion to resin bonds as well as improving its water resistance. Ceramic coating also provides the grain with a superior surface for better mechanical bonding in a wheel matrix. Heat-treating the abrasive and quenching the grain can increase friability. Alloying elements also adjust the friability of the grain; zirconium is a common alloy used in this way. Compounds such as chromium oxide and vanadium oxide are also added. These oxides mod-

Figure 2.3 A piece of aluminum oxide broken from the cooled charge.

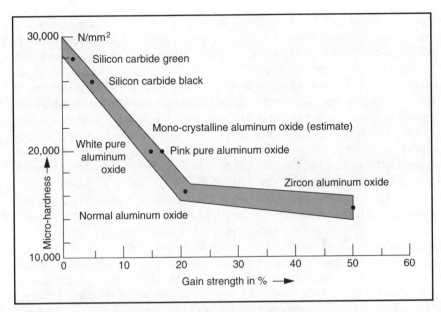

Figure 2.4 Properties of aluminum oxide and silicon carbide.

ify the toughness of the grain to a very small degree. However, it is said that the coloring effect, pink and blue-green, respectively, is used more as a marketing tool than for technical reasons (see Fig. 2.4).

Aluminum oxide is used in over 70 percent of the abrasive machining applications in the world. It is by far the most widely used of all abrasive media. The demand for large quantities of aluminum oxide has prompted more modern manufacturers of the grain to use a continuous fusion process. Here a furnace, similar to the one described above, is tipped every three to four hours. Depending on the property of the required grain, the molten aluminum oxide might be poured into large refractory molds and allowed to cool slowly to produce large crystal grains. Conversely, the molten aluminum oxide might be poured into water which will quench the compound and yield a very fine crystalline structure. Once cooled, the grain is further crushed and then washed prior to grading for size. The grain might then be postprocessed or coated prior to being combined with the bonding agent and formed into the shape of a grinding wheel.

2.4 Aluminum Oxide (Ceramic)

- Chemical formula—Al_2O_3.
- Standard wheel designation symbol SG.

Ceramic abrasive is the generic term which has been given to the aluminum oxide produced by the Sol-Gel process. There are apparently two types of such abrasive mineral on the market; "SG" (which stands for Seeded-Gel) manufactured by the Norton Company and "Cubitron" manufactured by the 3M Company. 3M has produced the Cubitron grain commercially since the late 1970s and used it exclusively in coated abrasive products. Norton launched their SG grain in vitrified bonded wheels in the late 1980s. Now both the SG and Cubitron grain is found in vitrified aluminum oxide products of the world's abrasive product manufacturers. The process used to manufacture the ceramic aluminum oxide grain produces a "solid" crystal of aluminum oxide with a great deal less friability than typically found in the fused aluminum oxide. However, the sharpness of the ceramic abrasive is much longer lasting than the traditional fused aluminum oxide. The Sol-Gel process is very much like growing crystals in a supersaturated solution and has the potential to allow the manufacturer to produce mineral grains to a given shape and aspect ratio, a feature never before found in an abrasive grain mineral. Perhaps one day we might find grinding wheels manufactured by orienting columnar grains so that their best cutting action and wear resistance is preferentially positioned for optimum performance. Presently, the Norton Company and others are combining the properties of the ceramic and fused aluminum oxide into vitrified grinding wheels for precision grinding. Pure SG wheels tended to produce high residual stresses in heat and crack sensitive materials so the pure SG wheels are generally used in rough grinding operations. The combination of the ceramic and the fused aluminum oxide tends to produce a wheel more suited to fine precision grinding (see Fig. 2.5). The wheel exhibits a high level of sharpness, an aggressive cutting action, good form holding, and long life. The properties of the SG grain are complemented by the friability of the fused aluminum oxide. The grade designation is typically written, for example, SG2 (20 percent alumina mix) SG5 (50 percent alumina mix). This is a very exciting technology in that a new, relatively low cost abrasive has inserted itself between the conventional abrasives and the very expensive superabrasives with measurable improvements in grinding by a factor of three to five times.

2.5 Cubic Boron Nitride (CBN)

- Standard wheel designation symbol B.

In 1950, the General Electric Company recognized the fact that there was an increasing need for a reliable source of diamond and went about the task of creating artificial diamonds. In 1954, GE had

Figure 2.5 Seeded-Gel (SG) aluminum oxide on the left and fused aluminum oxide on the right. (*Photograph courtesy of Norton Company.*)

successfully manufactured a synthetic diamond in the laboratory, and by 1957 introduced commercially available synthetic industrial diamonds.

As a result of work carried out in the development of synthetic diamonds, the General Electric Company introduced a new abrasive to the industry in 1969 with the trade name Borazon, or cubic boron nitride (CBN) as it is more commonly known. CBN is an abrasive with a hardness greater than that of silicon carbide and yet not as hard as diamond. CBN has the ability to machine even the hardest of steels to very precise forms and finishes with a great deal less wear than aluminum oxide or silicon carbide. Moreover, CBN is chemically inert in the presence of carbon. Unlike diamond, which is carbon based and has an affinity for the carbon in steels, CBN can machine steels at very high stock removal rates and with superior surface finishes. CBN does however appear to react with water at high temperatures. Water vapor dissolves the protective boron oxide layer leaving the surface exposed to hydrolysis and leading to rapid wear and therefore a low G-ratio. G-ratio is the term which describes wheel wear. A high G-ratio is generally felt to be beneficial as the grinding wheel wears less than a grinding wheel with a low G-ratio. A G-ratio number of, say, 10 means that the grinding wheel machines 10 times as much material as is lost from the grinding wheel.

Peculiarly, the G-ratio number has never included the amount dressed from the grinding wheel in preparation. In order to enjoy the high G-ratios promised by the manufacturers of the new abrasive, it was necessary in the early days to use a straight oil cutting fluid, as opposed to the more accepted water-based cutting fluids. Today, water-based cutting fluids have improved immensely, such that the differences in G-ratio between oil and water when grinding with CBN are not nearly so pronounced.

There are two major suppliers of CBN abrasive: GE Superabrasives and DeBeers. Originally each company had its own trade name: Borazon, CBN, and Amborite, ABN, respectively. Both companies have since agreed to call what is essentially the same mineral, "CBN."

CBN is manufactured by a process which requires very high temperatures and pressures. It is synthesized in crystal form from hexagonal boron nitride which is commonly called "white graphite," derived from the pyrolysis of boron chloride-ammonia ($BCl_3 \cdot NH_3$). The hexagonal boron nitride, composed of atoms of boron and nitrogen, is combined with a catalyst like metallic lithium in an environment in the range of 1650°C (3000°F), and pressures of up to 68,000 bar (1,000,000 psi). A black, opaque grain results. De Beers used lithium nitride as their catalyst and found that a yellow, translucent grain results, hence, their designation Amber Boron Nitride (ABN). The intense heat and pressure in the presence of the catalyst causes the nitrogen atom to donate an electron to a boron atom which then forms a chemical bond to the nitrogen atom and forms a very strong crystalline structure similar to that of diamond. The crystals are blocky in shape with noticeably sharp edges and corners. Until the mid-1970s CBN was manufactured in only two forms: Type I (GE) or ABN 300 (De Beers), uncoated and used in plated wheels, and Type II (GE) or ABN 360 (De Beers), consisting of 40 percent by weight uncoated and 60 percent nickel coated. The nickel coating provides increased bond strength for use in resin bonds (see Fig. 2.6).

The CBN referred to above has a grain structure which fractures along cleavage planes. The grain therefore wears until the force on the grain is sufficient to cause shearing of the cleavage plane to expose a sharper, keener cutting edge. This shearing force is very high due to the high hardness and toughness of the abrasive. Diamond truing is therefore the best method for forming or truing very fine detail in the grain on the periphery of a grinding wheel. Diamond has the hardness, and therefore the ability, to fashion the shape of the surface grains. The advent of reliable vitrified bond systems for superabrasives makes crush dressing practical, yet for the finest detail, diamond dressing systems are superior.

A new type of CBN grain structure has been developed by both GE Superabrasives and De Beers. The new grains are types 500, 510, 550,

Figure 2.6 Nickel coated CBN 560 (top) with CBN 550, an uncoated abrasive (below). Both are 40/50 U.S. mesh. (*Photograph courtesy of GE Superabrasives.*)

560, and 570 CBN by GE and ABN 600, 610, and 660 by De Beers. This new structure is microcrystalline. The grains tends to be quite large, as large as FEPA B851 (20/30 U.S. mesh), and appear to sharpen themselves in-process by very fine crystallographic breakdown. This attritious wear mechanism therefore exposes an aggressive surface as opposed to a worn flat surface typical in monocrystal grain. CBN grinding wheels manufactured with the microcrystalline grains should be expected to require less truing and dressing.

Compared with aluminum oxide and silicon carbide, CBN has minimal wear and therefore stays sharper for a longer period of time between dressing. An individual CBN grain will exhibit a life 100 times that of an aluminum oxide grain (see Fig. 2.7). Because of its inherent sharpness CBN tends to machine cooler, providing high surface integrity and superior surface finish. The cost of the abrasive is high; however, the benefits derived from the dramatically reduced amount of wheel dressing and the quality of the workpiece surface may be advantageous. It is important to appreciate that CBN requires a very rigid machine tool with the correct truing and dressing method employed. Machine tools designed along the more traditional lines of a grinding machine do not have the vibrational stability nor the high speed capability required for the economical use of CBN. CBN has been shown to perform more successfully with increased wheel speed, providing high stock removal rates and minimal wheel wear. High-speed grinding tests, using plated CBN wheels, have been carried out in excess of 300 ms^{-1} (60,000 sfm) and have shown longer wheel life and better quality workpiece surfaces. There are machines working in production in the United States using plated CBN wheels, running in excess of 150 ms^{-1} (30,000 sfm). At normal speeds there is one disadvantage with CBN in that it does not machine very soft or gummy materials easily. The plated grinding wheels clog and load up with the soft, sticky material. At ultrahigh speed however, it would appear that the chip formation changes and brittle fracture occurs in the softer materials, such that at high speed it would appear that both hard and soft materials will machine alike. The demands on the machine tool for dynamic stability under high-speed conditions are great. CBN, along with diamond, has been termed a "Superabrasive," so similarly it requires "super" machine tools which have been designed with the specific requirements of CBN and diamond abrasives in mind. The truing and dressing methods for these abrasives require high dynamic rigidity of the machine tool as well as the truing and dressing mechanism in order to cope with the inherent vibration experienced with superabrasives. Dynamic wheel balancing is essential.

2.6 Diamond

- Chemical formula—C.
- Standard wheel designation symbol D.

In the 1950s, both the General Electric Company in the United States and ASEA in Sweden independently discovered a synthesis process which created a crystalline diamond from graphite. In 1957, the General Electric Company announced that the commercial manu-

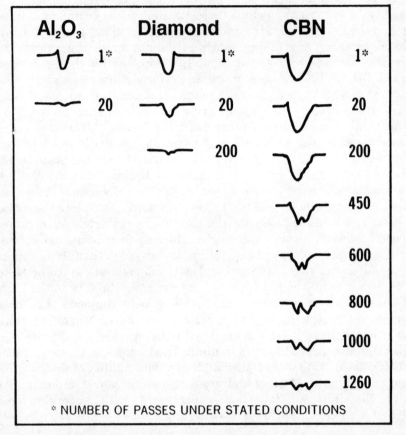

Figure 2.7 Grain wear.

facture of diamond, for sale as a consumable abrasive, was possible. The synthetic diamonds were given the trade name "Man-Made." The first successful industrially manufactured diamond was achieved when iron sulfide, in a graphite tube closed with tantalum end disks, was subjected to a pressure of 95,000 bar (1,400,000 psi) and 1600°C (2900°F) for several minutes. Tests on the crystal produced proved that it was indeed a diamond.

The industrial process for manufacturing diamond now uses pressures in the range of 55,000 to 130,000 bar (808,000 to 1,900,000 psi) at temperatures in the range of 1400 to 2500°C (500 to 4500°F). The catalyst-solvent metal interface is most important. Iron was first used, and since then chromium, cobalt, magnesium, nickel, platinum, rhodium, ruthenium, and tantalum have been used successfully. Different temperatures, solvents, and pressures produce different diamond types. Each crystal may be tailored to the best possible combination of size, shape, surface, and crystal structure for specific applications.

The metallic coating of the diamonds with nickel and/or copper provides better mechanical bonding in a variety of bond systems, as well as providing a path for the heat from the process to be conducted away from the diamonds, particularly in dry grinding applications.

Diamond is suited to grinding tungsten carbide, natural stones, granite, and concrete, as well as more sophisticated ceramics and cermets. Diamond, however, is most unsuitable for the grinding of steels due to the very aggressive chip formation which tends to tear the diamonds from their bond. Also, it is postulated that diamond, being a carbon-based material, has an affinity for the carbon in the steel and suffers accelerated wear by the dissolution of the diamond into the carbon in the steel, producing an iron carbide (Fe_3C) with most unsatisfactory results. These are two of the prime reasons for the introduction of CBN, which is less reactive in the presence of carbon steel alloys and has better mechanical bonding properties, making wheel fabrication that much easier.

2.7 The Properties of Abrasives

For an abrasive to function properly it must be harder than the material being machined. The abrasive particle must be of sufficient hardness and shape to be able to penetrate the surface of the material to be machined and form a chip or particulate. However, the abrasive must be tough enough to withstand both the thermal and mechanical shock of grinding, and at the same time be friable enough to fracture and produce new and sharp cutting edges.

Hardness is a term used in abrasive machining which is often confused. Hardness is usually referred to as a property of the grinding wheel. This will be discussed later. It must therefore be emphasized that hardness, as referred to in this chapter, is the property of the abrasive grain alone. The relative hardness of some popular abrasives is shown on this Knoop Scale:

Hardened Steel Rc 60	740
Quartz	820

Aluminum Oxide	2100
Silicon Carbide	2480
Cubic Boron Nitride	4700
Diamond	7000

Both mechanical and thermal shock resistance are the most important properties of an abrasive. The working abrasive grain endures not only intermittent cutting, but also thermal cycling. The heat of grinding may be detrimental to the abrasive in the presence of certain chemical elements, which could dramatically reduce the sharpness and hardness of the grain by diffusion into the grain matrix at high temperatures. This can be put into perspective when we consider that diamond is the hardest substance that we have discovered, and we assume that it is virtually indestructible. This is not a sound assumption because the hardness was measured by indentation hardness measurement. There was no sliding wear, frictional heat generation, or cyclical mechanical shocks. Hardness does not necessarily mean that the substance has good wear resistance. Diamond conducts heat tremendously well, six times better than copper and with very low thermal expansion. Though the diamond will not expand with increasing temperature, inclusions in the diamond will expand at a rapid rate and destroy the grain. Diamond quality is therefore an important area of concern when specifying natural or synthetic diamonds.

The need for toughness and friability in an abrasive seems to be contradictory. However, an abrasive must possess both qualities to a certain degree. In the extreme, a tough grain with little friability will become dull quickly and the grinding wheel will glaze. Dressing this type of grinding wheel with a single point diamond will tend to pull the abrasive grain from the bond, instead of breaking the grain to leave a sharp edge. Going to the other extreme, a friable grain with little toughness has a very aggressive nature, and will crack and fracture very quickly under load, revealing another layer of sharp-edged grains. This rapid shattering of the friable grain results in a cool cut, but at the expense of high wheel usage.

To reinforce the concept of toughness and friability it is worth considering glass as a material. Glass, which is significantly harder than a piece of mild steel, is a candidate for a tool material to machine steel, but glass has little toughness. If a drill or a milling cutter were able to be made from glass, we would see that the glass would shatter under the machining forces. If the glass were to get hot and then be quenched by the cutting fluid, again it would shatter into pieces. Glass is therefore too friable and lacks toughness. Toughness is the ability to hold shape without compromise in hardness.

There are a number of abrasive types as well as forms. Our choice of abrasive will be governed by a number of factors, all relating to the overall economics of the machining process, along with the type of material being machined, the surface finish required, dimensional accuracy, profile detail, and form tolerance. Particularly in the case of precision grinding, the type of machine tool being used and its physical condition also will influence the choice of an abrasive system.

When an abrasive type is first selected, it is graded into grain sizes. The system used to do this is a series of fine mesh wire grids (see Fig. 2.8). The mesh size corresponds to the number of openings per linear inch in a wire gauze. The gauze categorization is carried out for sizes 4 to 240. In many applications where fine finishes are required, the abrasive is categorized into much finer mesh sizes, which are too fine to be segregated by gauzes. The fine grain may be as fine as 4000 on the mesh scale. Wheels of such fine grain are sometimes referred to as "flour" wheels. Grains finer than 240 are separated by a flotation method where the grain is suspended in water. At given time intervals, the grain, which has settled, is extracted and as the settling time progresses, the grain is graded finer and finer. The "flour" grain is so fine that it will float on the surface of the water, held afloat by the surface tension.

Once the grain has been manufactured, processed, and graded into sizes, it is then bonded into a tool. The bonding system is the method by which the grains are held together in the shape of a grinding tool. If the grain is to be bonded into the shape of a precision grinding wheel, then the most common bonding systems are vitrified, resinoid, rubber, metal and plated. Approximately 50 percent of all grinding wheels manufactured are vitrified. If the grain is to be used as a coated abrasive then it will typically be bonded to a paper or a cloth backing.

2.8 Concentration

When a superabrasive grinding wheel is made there is an additional factor called concentration, which appears in no other abrasive system. It refers to the amount of abrasive per unit volume of usable wheel. Concentration numbers are typically in the range 30 to 175, 30 having only 1.32 carats/cm^3 (22 carats/in^3) and 175 having 7.7 carats/cm^3 (126 carats/in^3). The grain size will affect the achievable concentration level, as it is easier to completely fill a space with many smaller particles as opposed to very large randomly shaped blocks (see Fig. 2.9).

The concentration of the selected grinding wheel is based upon the area of contact between the wheel and workpiece. A large area dictates a low concentration and a small area a high concentration. A rule of thumb to establish a concentration number is as follows:

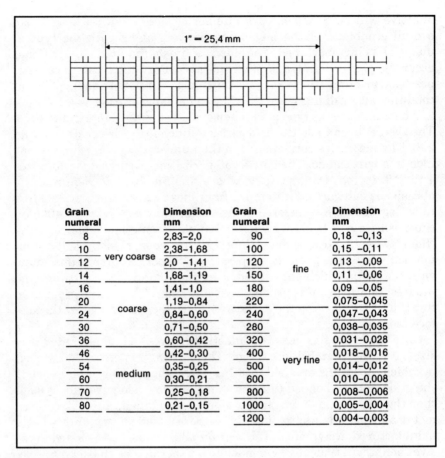

Figure 2.8 Grain size.

If the contact area between the grinding wheel and the workpiece is less than 15 mm² (0.025 in²), then use 100 to 150 concentration; between 15 and 50 mm² (0.025 to 0.075 in²), then use a 75 concentration; and for larger areas use 50 and maybe as low as 30 concentration, particularly for materials like glass which are prone to chipping.

Unfortunately, the industry sees concentration as more of an economic factor than a technical one. Confusion also arises with respect to the meaning of 100 concentration which is often misinterpreted as 100 percent and the maximum concentration. This is *not* so. A concentration of 100 means that there are 4.4 carats of abrasive per cm³ (72 carats per in³).

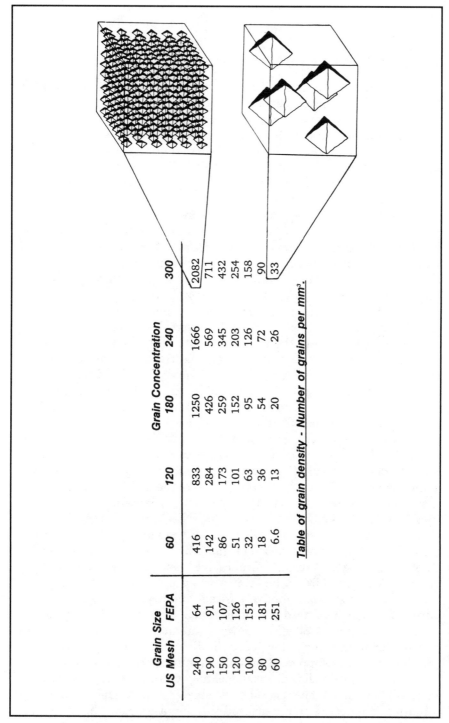

Figure 2.9 Grain concentration.

2.9 Vitrified-Bonded Grinding Wheels

- Standard wheel designation symbol V.

A vitrified bond is made of clay or feldspar which fuses at a high temperature to form a glass-like structure. During the firing operation, the clay or feldspar melts, surrounding the abrasive grain, bonding each grain to the next, and forming a homogeneous structure. When the wheel cools, each grain is surrounded by a hard glass-like bond which has high strength and rigidity. This type of bond system is well suited to abrasive machining as it fractures readily when the grinding forces build up on the abrasive grain. The grinding forces will increase on the grain as it dulls; generally, it is the dull grains that we wish to be broken free from the wheel in order to expose the keener, sharper cutting grains. The amount of bond material mixed in with the grain prior to firing will determine the strength of the grinding wheel and determine the grade of the wheel with respect to the hardness of the grinding wheel. The bond strength is the holding power of the bond to hold a grain in position under the grinding forces. In this instance hardness refers to the overall hardness of the grinding wheel, composed of the grain held into a bonding system. A balance may therefore be struck between the friability and toughness of the grain versus the strength and brittleness of the bonding agent. In one case the grain might be the first to fracture, while in another case the bond might be weak and the whole grain might be plucked or broken from the wheel periphery.

Grinding wheel hardness is determined by the grinding wheel manufacturer. One method of wheel hardness measurement is to take a hard spade-type drill with a constant thrust force and literally drill the grinding wheel. The depth of penetration of the drill is measured after a set period of time. The depth of penetration determines the wheel hardness and is the basis for a grinding wheel grade in the range of A through Z, with A being the softest and Z the hardest. The deeper the penetration over a given period of time, the softer the grinding wheel. Another method is to use an air/abrasive blast to break the grain from the bond system. After a blast at a given pressure for a given time using a known size of abrasive particle, the depth of erosion is measured and the wheel hardness determined. One other method utilizes a natural frequency of vibration measurement technique, a system called "Grind-O-Sonic," developed in Belgium (see Fig. 2.10). The grinding wheel is supported on four equally spaced points on an isolated rubber pad, such that the wheel may vibrate when given a sharp blow with a hard rubber hammer or similar object. The frequency of vibration is detected and measured through a pickup. The numerical value is entered into a formula which relates the size, shape, and mass

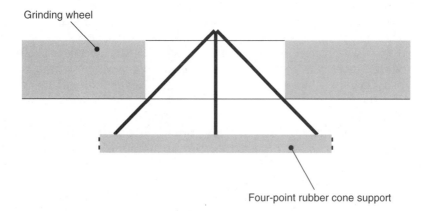

Figure 2.10 Grind-O-Sonic.

of the grinding wheel to the frequency of vibration, and determines a bulk Young's modulus for the grinding wheel. The Young's modulus can then be related with surprising accuracy to the hardness and performance of the grinding wheel. This technique has proved to be a major contributor to hardness "balancing" of rotary honing stones (see page 192).

A further property of a grinding wheel is its structure (see Fig. 2.11). The structure refers to the skeletal structure of the bond system. The structure is a measure of the density/porosity of the grinding wheel. Supposing a great deal of very fine abrasive grain were mixed with an equal amount of very strong bond material and pressed under high pressure, a dense, low porosity grinding wheel would result. If a small amount of grain were mixed with a small amount of bond material and another media (to space the grains apart), the result, once the spacing media were removed, would be a very open, highly porous structure grinding wheel. The latter method is used to manufacture the high porosity grinding wheels necessary for creep-feed grinding. The spacing media used to create the large and consistent porosity is paradichlorobenzene (moth ball crystals), which is removed from the wheel in its green state in a steam autoclave prior to firing. In the past, many different materials were used to increase the porosity of a grinding wheel; sawdust and walnut shells were quite common and usually left in the mix to burn out during the firing operation.

A vitrified grinding wheel is manufactured by selecting the correct abrasive and grain size, and thoroughly mixing the abrasive with the correct amount of bonding agent and porosity media, along with a little water. The mix is then packed and pressed into a grinding wheel mold, with pressures varying from 10 to 675 bar (150 to 10,000 psi). The mold is then fully dried, forming a grinding wheel in a green state. Shaping or recessing the grinding wheel by machining is more easily performed in the green state. If there is a pore inducing media in the mix it is removed in a steam autoclave. The wheel is then dried and fired in a kiln, in a similar manner to firing a piece of pottery, at temperatures approaching 1400°C (2500°F) for several days, depending on the size of the grinding wheels and the charge. The wheels are then removed from the kiln and slowly cooled. They are then checked for distortion, shape, and size. After machining to a final size, the wheels are balance tested, balanced, and overspeed tested, generally at 1.5 times the rated Maximum Operating Speed (MOS), to ensure operational safety.

An alternative to pressing the grinding wheel mix to form a wheel in the green state is a method called puddling. A puddled wheel is typically mixed to such a consistency that the mixture can be poured into a shaped mold and allowed to set before firing in the kiln. This method of wheel manufacture allows a larger and more consistent porosity throughout the grinding wheel, particularly across the wheel width, where pressing tends to develop a wheel much harder at the edges than in the center. This method usually results in a very open and soft structure suitable for creep-feed grinding.

It should be understood that the vitrified bond system is hard and brittle. The great majority of wheel wear takes place by the mechanical action of stressing the bond with a high grinding force, breaking the bond bridges, and allowing the exposure of a new, sharper, grain deeper in the wheel's structure.

There is an interesting note to be made with respect to silicon carbide and superabrasive wheels in a vitrified bond system as we have

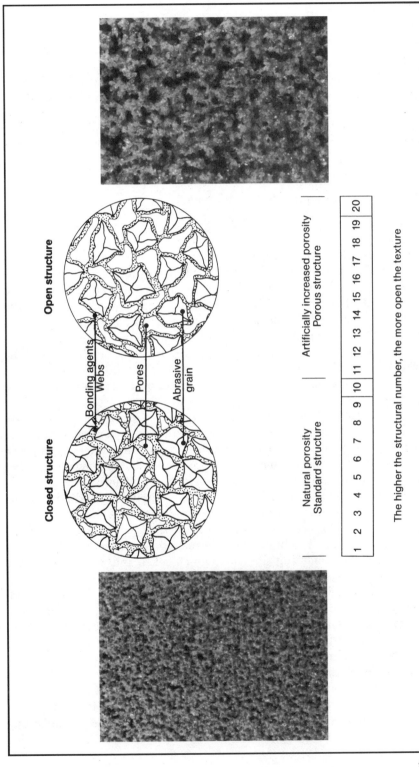

Figure 2.11 Grinding wheel structure.

described it. A silica, glass-like, vitrified bond media reacts adversely with a silicon carbide grain, so a porcelain/ceramic type bond system has to be used for silicon carbide wheels. Superabrasive grain can be bonded in a vitrified bond. However, diamond turns into graphite at 700°C (1300°F) and CBN begins to oxidize at 1000°C (1850°F) and completely oxidizes at 1900°C (3500°F). Therefore, lower temperature vitrified bonding systems had to be developed; remember, vitrification of Al_2O_3 wheels takes place at 1400°C (2500°F) (see Fig. 2.12).

2.10 Resin-Bonded Grinding Wheels

- Standard wheel designation symbol B.

Resin-bonded wheels are manufactured in a very similar manner to the vitrified wheels. However, the bonding medium is a resin. A thermosetting synthetic resin is mixed in either powdered or liquid form (latex) with the abrasive grain and a plasticizer (catalyst) to allow the mixture to be molded. The mixture is then pressed and cured at a temperature of 150 to 200°C (300 to 400°F) for periods of as little as twelve hours and as long as four to five days, depending on the size of the wheel. During this curing, the mold first softens and then hardens as the oven reaches curing temperature. Upon cooling, the mold retains its cured hardness.

Two disadvantages of resin-bonded wheels are their low porosity and, when in the presence of a cutting fluid, their tendency to soften and wear excessively. Much research is being carried out to improve the compatibility between cutting fluids and resin bond systems. Their prime areas of use are for high-speed grinding, where they can withstand much higher bursting forces than vitrified grinding wheels. Special resin-bonded wheels have been safely run at speeds up to 125 ms^{-1} (25,000 sfm), whereas vitrified wheels tend to be safe only up to 60 ms^{-1} (12,500 sfm). The limitation of a resin bond at very high speeds is the result of overheating the bond, which cokes and breaks out of the wheel periphery. Another prime area of use for resin-bonded wheels is steel mill snagging operations, and hand grinders, which suffer rough handling and abuse. The amount of hand grinding and snagging in the industry, as well as high-speed, superabrasive, and precision grinding, means that resin-bonded wheels account for 30 to 40 percent of the grinding wheel market.

It should be understood that the resin bond system is somewhat soft and forgiving, and that the great majority of wheel wear takes place because of thermal action, which melts the bond with frictional heat, thus softening the bond and allowing the dull grains to become dislodged and eventually torn from the wheel periphery.

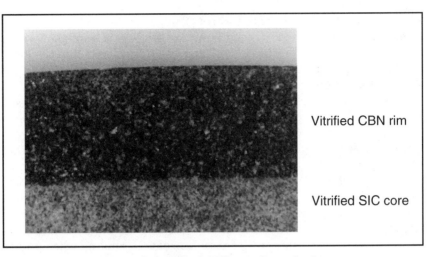

Figure 2.12 A rim section of a vitrified CBN grinding wheel.

2.11 Rubber-Bonded Grinding Wheels

- Standard wheel designation symbol R.

Rubber-bonded wheels are made by selecting the grain, sieving it like before, and then kneading the grain into a natural or synthetic rubber. Sulfur is added to the mix as a vulcanizing agent and then the mix is rolled between steel rollers to form a sheet of the desired thickness. The grinding wheels are then cut out of the rolled and sized sheet rather like cookies, using a cookie cutter. The wheels are then vulcanized under pressure at temperature of 175°C (350°F).

Rubber wheels claim a little less than 10 percent of the market. They can be made extremely thin, as thin as 0.050 mm (0.002 in). These very thin wheels are used for slitting fountain pen nibs. Conversely, very thick wheels can be produced for centerless grinding control wheels. Other applications for rubber-bonded grinding wheels are in the bearing industry, where extremely high surface finishes are required.

2.12 Metal-Bonded Grinding Wheels

- Standard wheel designation symbol M.

There are two divisions of metal-bonded wheels: those which have been plated, and those which have been cast. Included in the plating or cast matrix is the grain, usually CBN or diamond.

In the case of a plated grinding wheel, the wheel hub is manufactured very accurately with respect to the form profile and the profile's concentricity to the bore of the wheel. The profile is machined with an offset in the true shape, which allows for the size of a single layer of grain being plated. The abrasive is then carefully plated onto the form, using a hard, typically nickel or chrome plating to hold the grains in place. The plating does not completely cover the grains, and they are exposed with their sharpest cutting edges outermost. The thickness of the plating, a function of the grit size, corresponds to approximately 1 to 1.5 times the grit thickness. Most metal-plated wheels have a periphery of only one or two abrasive grains deep. Such wheels do not require dressing; however, they have to be very carefully assembled onto the machine spindle in order to run true to the spindle rotation. Most wheels of this type have a truing groove to assist in the proper mounting of the wheel. Wheel run out of less than 0.012 mm (0.0005 in) TIR is generally acceptable. Electroplated grinding wheels can run at very high speeds, since the central hub of the grinding wheel can be made from high-strength alloys.

A new technology has been developed called "Diamesh," which allows the superabrasive grain to be deposited and plated in a very tightly controlled pattern (see Fig. 2.13). There is no random placement of the grain, which is typical in conventionally plated wheels. Each grain is located in a mesh cavity to provide a regular pattern and even coverage of the grain across the wheel periphery. Diamesh provides a more consistent wheel performance, longer life, and better part surface integrity.

In the case of cast metal bond wheels, they are typically small grinding wheels used for Internal Diameter (ID) grinding and made from a soft bronze or other copper alloy, although there are some made from cast iron. The metal bonding is achieved using a sintering process and provides a very strong, solid bond, which has very good form retention properties. There is little to no porosity in a metal bond, but it can be enhanced somewhat by the addition of graphite fillers. Truing and dressing of these types of wheels is best performed by Electro-discharge Machining (EDM). The EDM process erodes the metal matrix, as well as what is usually a diamond abrasive, and allows relatively complex forms to be dressed into the wheel periphery. The EDM process can be sufficiently controlled to cause a pitting of the surface, which can act as a cavity for chip clearance and cutting fluid flow. Diamond is often used in this bond system, and finds a niche in the machining of hard ceramic materials and glass. Large wheels, above 150 mm (6 in) in diameter, can be manufactured in a sintered metal bond; however, the wheel is generally assembled from a series of separately sintered segments adhered to the wheel periphery. The separately sintered segments suffer from density changes and inconsistencies among the various segments. Such segmentation, and therefore separation, of the

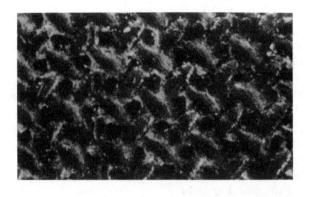

Figure 2.13 Type S/1 Diamesh which has a single abrasive particle per mesh opening (top) and type M/1 Diamesh with two abrasive particles per mesh opening (bottom). (*Diamesh is a trade name of UAS Inc.*)

wheel periphery can result in high vibration levels and poor surface integrity.

A crush dressable metal bond has been developed, which, by its name, suggests that it can be dressed to a given form profile by a crushing action, but is actually more like a compressing action of a very porous metal matrix.

The bonding and structure of a grinding wheel determine its safe operating speed. The safe speed is printed on every grinding wheel and must never, under any circumstances, be exceeded. Every grinding wheel is tested for safe operation to 1.5 times the Maximum Operating Speed (MOS). Should a grinding wheel burst, there is a high risk of very badly wrecking the machine, the fixture, and the workpiece, and quite possibly injuring the operator. Be very careful, treat the process with respect, and never over-speed a grinding wheel. Maximum speeds are often printed in RPM. However, in new machine tool designs and control systems, wheel speeds are controlled to maintain a constant peripheral speed. This means that as the grinding wheel gets smaller in diameter, the RPM has to increase in order to maintain a constant surface speed. Therefore, the rule is that the maximum operating RPM is with respect to the largest wheel diameter when the wheel is new. It is fast becoming the norm to see both maximum RPM and maximum peripheral speed (MOS in ms^{-1} or sfm) marked on the wheel.

Not only should proper care be taken to operate a grinding wheel, but also safe mounting, handling, and storage of grinding wheels should be most important. The American National Standards Institute Safety Code, ANSI B 7.1, outlines the recommended methods for storage and handling (see Fig. 2.14).

2.13 Abrasive Belts—Coated Abrasives

The term coated abrasives refers to abrasive grain which has been adhered to a backing. There are three components to a coated abrasive product: the abrasive grain, the adhesive or coat, and the backing material (see Fig. 2.15).

All of the abrasives discussed previously are used in the manufacture of coated abrasive products. Aluminum oxide is the most popular and is used in the majority of coated abrasive applications. The abrasive is hard and durable, as well as being low in cost. Aluminum oxide is a tough abrasive and lends itself to applications that require heavy pressure. Grinding high tensile strength steel, metals, and hard woods are the prime reasons for using aluminum oxide.

Garnet is a semiprecious stone formed from a natural spinel. A spinel is a chemical crystal formation of a metal and aluminum oxide, e.g., $FeO \cdot Al_2O_3$ and $MgO \cdot Al_2O_3$. It is deep red in color and is used as a natural abrasive, in particular by the woodworking industry. It is often heat treated to increase its friability and improve its cutting ability in coated abrasive applications.

A new abrasive which is an agglomerated grain has been created by the 3M Company. Small grains of abrasive have been bonded together to make a grain somewhat like the microcrystalline CBN. Such an agglomerated grain will break down under heavy load and give the abrasive belt a property much like the self-dressing capability of a vitrified bonded grinding wheel.

The coated abrasive is selected according to the grain size which will yield the required surface finish. Both aluminum oxide and silicon carbide are available in grain sizes from 12 to 600. However, very intricate and detailed forms cannot be machined using coated abrasives. The coated abrasive belt can be made to conform to a shoe or formed platen, allowing the finishing of contoured parts, such as golf club heads, water faucet spouts, and surgical implants.

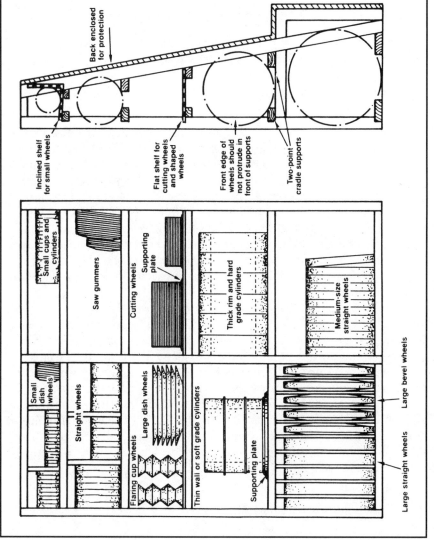

Figure 2.14 Methods for storing grinding wheels.

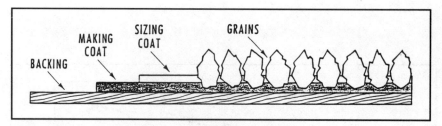

Figure 2.15 Cross section of a coated abrasive product.

2.14 Backing Materials for Coated Abrasives

Having selected an abrasive type and size, the next choice is backing. Coated abrasive backings are made from paper or cloth. Paper backings are the least costly and come in a variety of weights. The weight is measured by the weight of a ream (480) of 24- by 36-in sheets and given a letter code: A, B, C, D or E. "A" weight is 18 kg (40 lb), "B" weight is 23 kg (50 lb), "C" weight is 30 kg (60 to 70 lb), "D" weight is 45 kg (90 to 100 lb), and "E" weight is 60 kg (130 lb). Paper belts are nearly always "E" weight; lighter papers are termed cabinet papers. Paper backings are used when pliability and strength of the backing are not important. Paper-coated abrasives are most commonly used in the woodworking industry.

Cloth backings fall into a number of categories. There are two twill weave cloths: "drills," marked with an "X," and "jeans," marked with a "J." The main differences are that drills are made from heavier threads with fewer threads per square inch than jeans. Drills are the stronger of the two and used primarily with coarse abrasives for heavy work. Jeans are more flexible and ideally suited to finishing type operations.

The key factor in the selection of a cloth belt is its tearing limit under pressure. Cloth belts are pushed to their limit with today's demands of cutting speeds and stock removal rates. Synthetic belts are now being manufactured with complex weave patterns, which provide directional strength and longevity.

2.15 Adhesives for Coated Abrasives

The choice of adhesive is dependent upon the application. If the application is dry, then a straight glue (animal hide) can be used. If a cutting fluid is used, then a liquid phenolic resin is used. A balance has to be struck between the curing times and properties of the adhesives. Depending on the adhesive, some may soak into the backing and cause the belt to become stiff or too lightly coated, so that the abrasive is easily torn or flexed from the backing in operation.

2.16 The Manufacture of Coated Abrasives

The key step in coated abrasive manufacture is the application of the abrasive. This may be done by pouring the abrasive in a controlled stream onto the adhesive-impregnated backing, or more commonly running the impregnated backing through a tray of abrasive and allowing it to pick up the grain. Electrostatic attraction and orientation is also used to cause the grain to imbed itself firmly into the adhesive, oriented "sharpest point uppermost," resulting in a very aggressive and fast cutting tool.

The pattern and amount of abrasive being placed onto the backing can be very closely controlled to provide two types of abrasive systems: an open coating (covering 50 to 70 percent of the surface), which has spaces between the abrasive grain for chip clearance, or a closed coating, where there are virtually no spaces between the grains. Open coating is best suited for more flexible applications and closed coated for very arduous conditions. Once the grain has been deposited, the belt is coated with the final "sizing" coat to anchor the abrasive onto the backing, and then rolled into a giant roll for storage.

Before the coated product is used, it has to be processed into a marketable form. The adhesive will cause the belt to become very stiff and has to be "broken" in a controlled manner, usually perpendicular to the edge of the belt. The break lines are determined by the grain size of the

Figure 2.16 Lap joint in a 100 mm (4 in) wide abrasive belt.

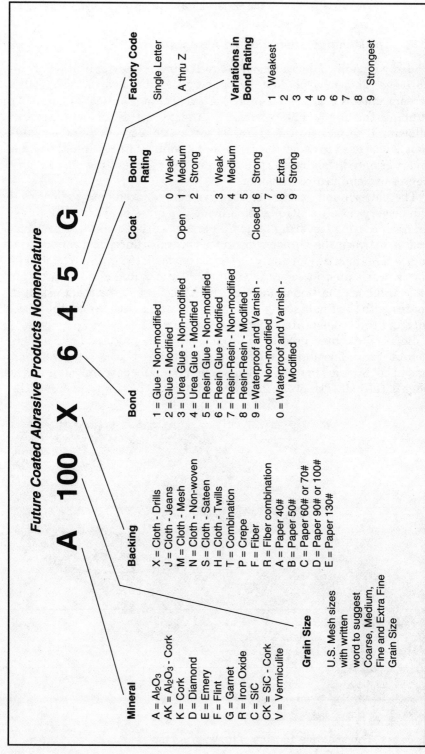

Figure 2.17 Coated abrasive designation system.

abrasive—closer for fine grain and wider for coarse grain. Other break line patterns are used for a variety of applications. Too much flexing and breaking of the bonding reduces the life of the belt and, therefore, is kept to a minimum.

Papers and sheets are cut from the large production rolls of coated abrasives and packed. Belts have to be cut and joined. The joint is a lap joint (see Fig. 2.16), generally 45 degrees to the edge of the belt. However, for narrow belts the angle is usually more acute, and for wider belts more obtuse. To form the lap joint without a significant lump in the belt, one end of the belt has the abrasive removed from the backing and the other end has a very small amount of the backing removed, so that when the joint is made, it is virtually invisible and will not upset the surface finish of the workpiece being machined. Coarse grain belts, which are not expected to produce high quality surface finishes, are usually lap jointed by removing the abrasive from only one end of the belt. The lump in the belt will not effect the resultant, rough surface finish.

Unlike the system for grinding wheels, there is no standard nomenclature for coated abrasives. However, there is a proposed unofficial system with similar information (see Fig. 2.17).

2.17 The Storage of Coated Abrasives

The backings and adhesives used in coated abrasives are very sensitive to climatic variations in temperature and humidity; resin bonded grinding wheels suffer the same fate. It is therefore most important to store coated abrasive and resinoid products in an environment which reduces their degradation by atmospheric conditions. The ideal environment is a temperature between 16 and 24°C (60 to 75°F) and relative humidity between 35 and 50 percent. Abrasive belts, in particular, should be hung in an attitude similar to the curvature of the wheels on which they will be used, in order to prevent undulations in the belt, which may upset the machined surface finish.

Chapter 3

Abrasive Preparation

3.1 Grinding Wheel Preparation—Mounting

The handling and mounting of vitrified grinding wheels, in particular, has to be carried out with the utmost care and attention. A grinding wheel rotating at normal working speed, 30 ms^{-1} (6000 sfm), has a peripheral speed in the region of 115 km/h (70 mph), and is a potential source of disaster unless the strictest practices are adhered to. The grinding wheel must be checked for soundness prior to mounting. Too often grinding wheels are abused in handling and shipping, or stored in poor conditions. Therefore, it is essential to "ring" a vitrified grinding wheel to be sure there are no hairline cracks in the wheel, which could cause the wheel to burst and break while in use.

Ringing is a method whereby the grinding wheel is held delicately in the bore and, using a small wooden mallet or similar object, struck gently, but firmly, while listening for a clear ringing sound as if ringing a piece of porcelain china. A clear ringing sound indicates that the wheel has no cracks. The wheel should be rung four times (at 90 degree angles) to be sure that the wheel is crack-free. The grinding wheel should be mounted on the machine arbor using plastic or paper blotters between the grinding wheel and the metal flanges (see Fig. 3.1). Blotters provide the means to grip the sides of the grinding wheel and prevent the metal flanges from breaking into the sides of the grinding wheel and initiating a crack. Paper blotters have been the traditional choice, however, there have been occasions where it has been necessary to use plastic blotters for better lateral stability of the wheel. The paper blotter will absorb the cutting fluid and swell or become pliable and allow a very small movement of the wheel between the flanges, which would be undesirable during a cross-feed, stepping, slotting/form, grinding operation. The flanges should be matched so that clamping is evenly distributed across the face of the grinding wheel. The flange bolts should be tightened, in sequence, to achieve a uniform clamping force (see Fig. 3.2). The mounted grinding wheel should now be statically balanced (see Fig. 3.3). Once balanced, the balancing weights must be tightened. Next, the wheel should be secured to a clean spindle, and,

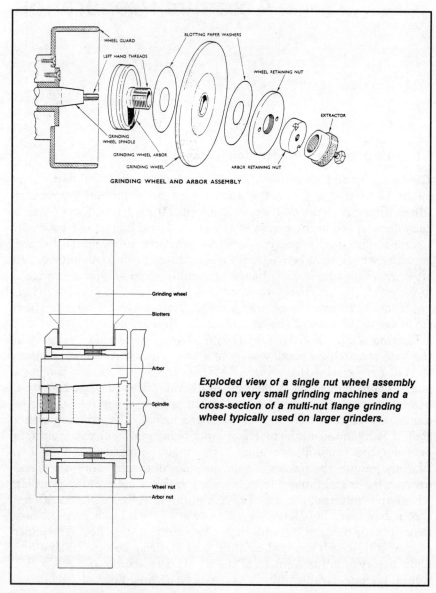

Exploded view of a single nut wheel assembly used on very small grinding machines and a cross-section of a multi-nut flange grinding wheel typically used on larger grinders.

Figure 3.1 Grinding wheel assembly.

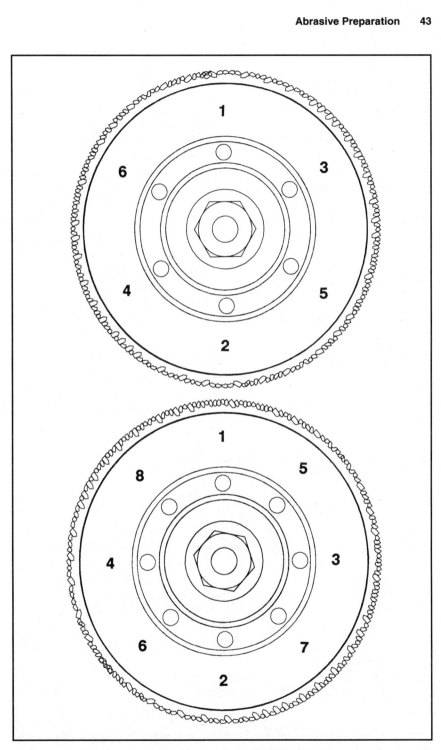

Figure 3.2 Bolt tightening sequence for six and eight bolt flanges.

where possible, slowly brought up to grinding speed. The grinding wheel should be dressed at the intended operating grinding speed until the complete periphery of the wheel has been swept by the diamond or dressing system. This can be verified by chalking a mark around the wheel periphery and observing when the mark disappears. The wheel should then be removed and statically balanced once more, and then remounted onto the spindle. After a final dress the grinding wheel is ready for use. There must be adequate guarding of the grinding wheel at all times to prevent damage and/or injury to the operator, casual passersby, or equipment. During the early stages of wheel mounting and balancing, wheel breakage is most likely to occur. Be careful! Should a break occur there is usually one of two breakage patterns. If the wheel breaks in two pieces, then it is most likely that the grinding wheel was improperly clamped between the flanges. If the wheel breaks into three pieces equally divided into 120 degree segments, then the wheel was probably oversped. This breakage pattern is due to the distribution of the bursting force exerted in the bore of the wheel.

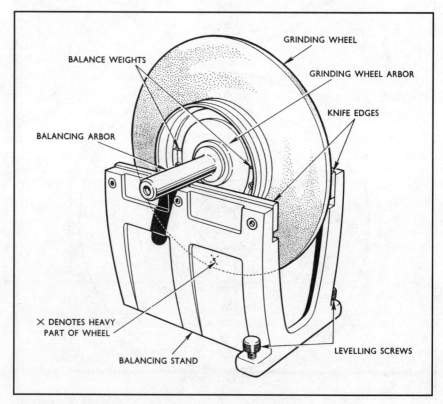

Figure 3.3 Balancing a grinding wheel on knife edges.

It can take a significant amount of time and lead to much operator frustration to balance a grinding wheel manually (see Fig. 3.4). There are units available today for on-machine balancing. A stroboscope/accelerometer system can assist in on-machine balancing. The stroboscope/accelerometer system detects balance relative to wheel speed, therefore the grinding wheel has to be running at operating speed to begin the balancing procedure. The stroboscope lamp flashes at wheel frequency in sequence with the accelerometer to show the minimum out-of-balance position, where a balancing mass should be placed. Though this is a semi-manual balancing system, it is quite accurate and can be quite fast. On more modern machines, auto balance units may be fitted to the rotating spindle as part of the wheel flange assembly. These units keep the grinding wheel in balance throughout the life of the grinding wheel, rebalancing the wheel each time it is stopped and restarted or upon request. One such system uses the principle of injecting cutting fluid into pockets which form part of the wheel flange (see Fig. 3.5). The mass of the fluid acts as weights balancing the grinding wheel while it is running. Other systems move weights mechanically and have been found to be more reliable, particularly at very high wheel speeds (see Fig. 3.6).

Probably one of the most hazardous situations associated with wheel mounting is the remounting of a grinding wheel, which has been used previously. Always check the bore of the wheel for nicks or cracks. In particular, be careful to spot where the blotter may have been trapped in the bore of the wheel. This could be a potential danger. Be sure to remove all blotter material from the side face of the grinding wheel. Some manufacturers adhere the blotter to the grinding wheel, which, if the blotter becomes torn or damaged, can present quite a challenge. A wheel burst is initiated at the center of the wheel, where the separating forces are the highest. Never force a wheel onto a spindle if the fit in the bore seems tight.

It is always good practice to allow a grinding wheel to spin dry for a period of time before unmounting, so that all of the cutting fluid contained in the wheel's porosity has a chance to spin clear. Unfortunately, the wheel you might be mounting may not have been spun dry or may have been left standing in one position for a significant amount of time, so that the fluid has collected in one place or has soaked up fluid from a source on the floor. Unbalanced fluid within the pores of a grinding wheel is one of the most common causes of start-up vibration in grinding machines, so insure that the wheel is dry before balancing and mounting.

Metal-bonded wheels are a special case, as they do not require blotters for mounting. A metal-bonded, usually superabrasive, wheel has a bore slightly larger than the spindle diameter to allow for concentricity

Chapter Three

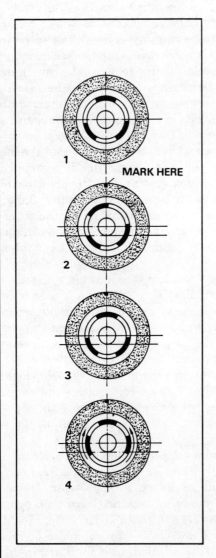

THREE WEIGHT WHEELBALANCING

1. Securely lock wheel onto flanges.
2. Position the three balance weight segments equidistant around the face of the flange.
3. Fit to wheelspindle and dress wheel.
4. Remove assembly from wheelhead and mount on balancing mandrel. Place assembly on the balancing unit, allow it to turn until it stops, and mark top centre of the wheel with chalk.
5. Move the segments equally around the flange until one segment is aligned with the mark.
6. If movement still occurs gradually move the other two segments equally towards the mark until the assembly remains static in any position.
7. Refit the assembly onto the wheelspindle and redress the wheel prior to grinding.

Figure 3.4 Manually balancing a grinding wheel.

TWO WEIGHT WHEEL BALANCING

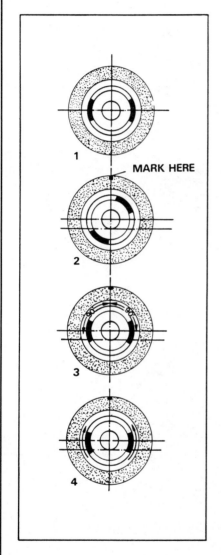

1. Securely lock wheel onto flanges.
2. Position the two balance weights diametrically opposite.
3. Fit to wheelspindle and dress wheel.
4. Remove assembly from wheelhead and mount on balancing mandrel. Place assembly on the balancing unit, allow it to turn until it stops, and mark top centre of the wheel with chalk.
5. Move the segments to 90° from the mark so that each weight is diametrically opposite.
6. Move weights equally towards the mark until assembly remains static in any position.
7. Refit the assembly onto the wheelspindle and redress the wheel prior to grinding.

Figure 3.4 *(Continued)*

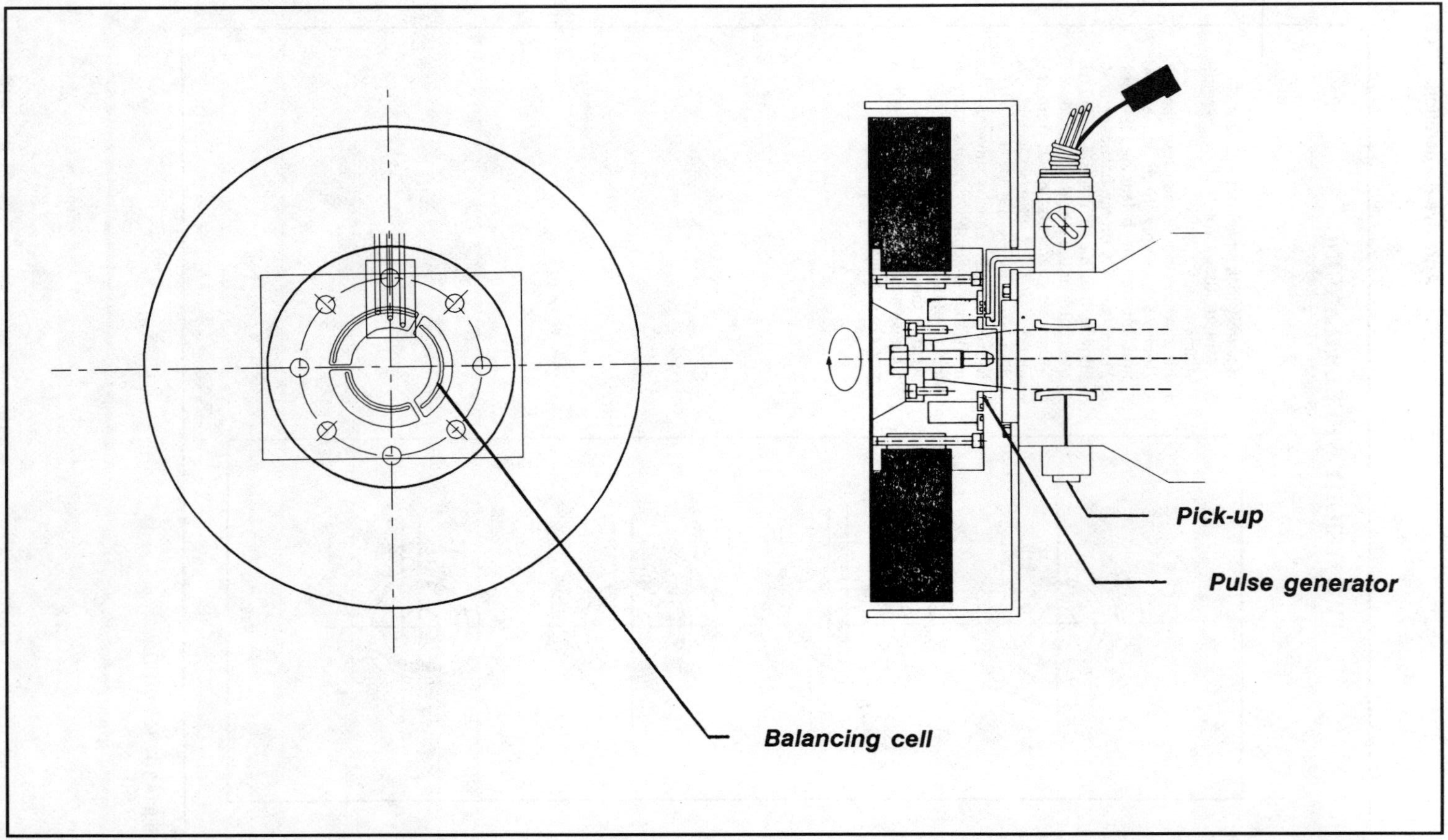

Figure 3.5 Fluid injection balancing. (*System by American Hoffman Corp.*)

Figure 3.6 Electro-mechanical dynamic balancing. (*System by SBS-Schmitt Industries.*)

adjustment. A single layer superabrasive wheel cannot be trued by a dressing operation. It has to be mounted to an indicating groove which directs the manual adjustment of the wheel between the flanges. This adjustment is carried out using a wooden block against the wheel periphery and a dial indicator in the indicating groove. The block is hit sharply with a mallet to offset the wheel, until concentricity, better than 0.012 mm (0.0005 in) TIR, is achieved (see Fig. 3.7). It has always to be assumed that the grinding wheel has been manufactured concentrically to the indicating groove. This is not always the case.

3.2 Fitting a Coated Abrasive Belt

The precautions for fitting an abrasive belt are not as rigorous as mounting a vitrified grinding wheel. However, there are certain procedures to be followed. Firstly, the contact wheel should be checked for any flat spots, nicks, or misshapes. The belt should be checked for cracks in the backing, tears, or poorly adhered abrasive. If the belt appears in good order it can be looped around the contact wheel and tension pulleys in the running direction, if indicated. The tension in the belt depends on the operation, and is typically as low as 0.875 N/mm (5 lb/in) belt width for light finishing work, and up to 7 N/mm (40 lb/in) belt width for roughing and high stock removal grinding. A good guide to belt tension is to tension the system so that during operation the belt does not "walk" across the contact wheel when the workpiece is applied. It is also good practice to keep at least 100 degrees of belt wrap around the contact wheel before the workpiece point of contact.

Abrasive belts operate at various speeds and have very much the same variability in performance as grinding wheels. Generally, they can operate at higher speeds than most grinding wheels, with a maximum around 50 ms^{-1} (10,000 sfm), and have the added advantage of very high-stock removal capability. High-speed and high-stock removal rates mean high-power drives. Typically the belts are rated from 0.4 kW to 7.5 kW (0.5 hp to 10 hp) per inch width of belt. Once the belt has been selected and fitted, all guards must be in position before the machine is switched on.

3.3 Grinding Wheel Conditioning

A grinding wheel is seldom ready for grinding once it has been mounted on the spindle of the machine. Of course, plated wheels are the exception. Careful preparation of the grinding wheel periphery has to be carried out in order to achieve the correct form, concentricity, and cutting action for the desired workpiece surface. The preparation involves two important steps—truing and dressing.

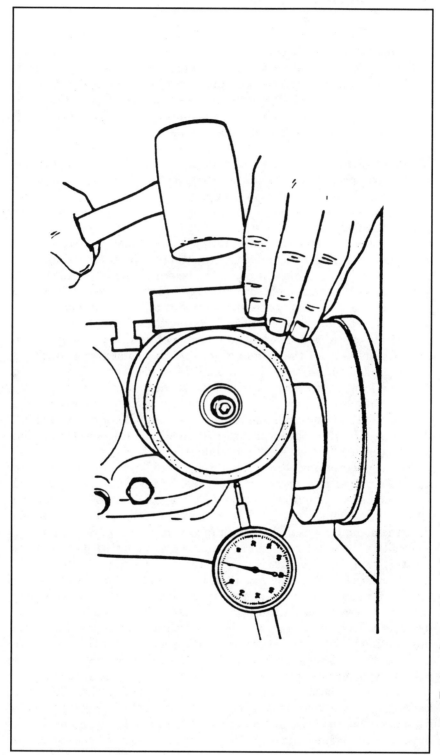

Figure 3.7 Truing of a metal-bonded wheel.

Truing a grinding wheel almost speaks for itself. It is the method by which a grinding wheel is mounted on a grinding spindle so that both the radial and axial run out is minimized or eliminated. Certain grinding wheels are never dressed (e.g., single-layer, plated, superabrasive wheels). These wheels can be trued only by accurate mounting. Superabrasive wheels in both resin and vitrified bonds are also mounted carefully to minimize their run out. These wheels have a peripherally thin layer of very expensive abrasive. Minimizing the run out prior to truing and dressing is driven by the abrasive cost and not necessarily the time to perform a lengthy truing operation.

Vitrified bonded wheels, be they conventional or superabrasive, are generally trued and dressed in one operation. This nomenclature has caused much confusion in the industry, as the combined truing and dressing of vitrified wheels is generally referred to as "dressing." The combined truing and dressing operation begins by mounting the grinding wheel within reasonable concentricity and then dressing the wheel periphery to assure exact concentricity, while at the same time imparting a level of peripheral sharpness. Dressing the grinding wheel can be carried out using a number of different methods, depending on the desired results from the grinding process. The combined truing and dressing operation not only conditions the wheel periphery, but also defines a form on the wheel periphery. The form may be a simple flat form for plain surface grinding. It may be a very complex form, requiring extremely tight profile tolerances, less than 0.005 mm (0.0002 in). Remember, dressing the form onto a vitrified grinding wheel trues and dresses the grinding wheel at the same time.

Truing and dressing become two distinctly different operations when using resin-bonded superabrasive wheels. One popular truing method is carried out using a brake dresser (see Fig. 3.8). A brake dresser is a rotary device which is limited to a slower peripheral running speed than the grinding wheel. An aluminum oxide or silicon carbide wheel, generally a 46 to 60 grit J to M hardness, is mounted on the brake dresser, brought in contact with the grinding wheel with a down-feed in the order of 0.02 to 0.04 mm (0.001 to 0.0015 in), and traversed at 0.5 to 1.5 m/min (20 to 60 in/min). It is important to engage the feed with the brake dresser turning. Otherwise, a flat can develop on the dressing wheel, inducing a vibration which will adversely affect the preparation of the wheel periphery. The brake truing operation results in the removal of material from both the brake dressing wheel and the resin-bonded grinding wheel. An allowance has to be made for the radial wear on both wheels, usually in the range 10:1 to 20:1 (the majority of the wear taking place on the brake dressing wheel), depending on the size and grade of the wheel used on the brake dresser. This means that the final wheel diameter has to be determined by touching off the workpiece or in-process sensing. This poses a particular disadvantage in CNC or automatic operations, where knowledge of the exact wheel dimensions is an inherent part of the grinding cycle. In addition, truing may be carried out using a metal-bonded, impregnated diamond nib to traverse across the wheel periphery, typically FEPA D126 to D301 (60/120 U.S. mesh). Wear on the diamond nib is generally minimal per dress cycle and provides an alternative for CNC and automatic operations. A single-point diamond must never be used to true a resin-bonded, superabrasive grinding wheel, as the diamond wears rapidly, generates an intense localized heat at the diamond tip, and causes thermal as well as mechanical destruction of the bonding layer.

Another and more efficient method of truing a resin-bonded wheel is to mount the resin-bonded wheel on a cylindrical grinding machine and use an aluminum oxide or silicon carbide grinding wheel to grind the resin-bonded wheel true. The aluminum oxide or silicon carbide grinding wheel should be run at 20 to 30 ms^{-1} (4000 to 6000 sfm) and the resin bonded grinding wheel at 0.5 to 2 ms^{-1} (100 to 400 sfm). The grain size of the dressing wheel should be between 15 to 30 percent larger than the grain of the wheel being trued. The grinding wheel may

Figure 3.8 A brake controlled dresser used to true a resin-bonded superabrasive wheel.

be very quickly trued by this method, but there is a disadvantage. This truing operation is not carried out on the grinding machine it is intended to be used on. Any error in mounting concentricity will therefore show up in the grinding wheel.

It is very important to recognize that these truing operations true the wheel *only,* causing the periphery of the wheel to be closed and *not* suitable for grinding. The wheel has to be subsequently dressed to prepare the periphery for grinding.

Dressing a resin-bonded wheel entails opening the bond to expose the grain by using an aluminum oxide stick to "stick" the wheel. The grain size of the stick should be FEPA D64 (230 to 270 U.S. Mesh) or finer and typically F to J hardness. Using hand pressure, the stick is pushed against the wheel periphery as it rotates at operating speed. "Sticking" is best performed wet, as it will keep the dust down, but moreover it helps to develop a slurry in the nip between the stick and the wheel periphery, and enhances the erosion of the resin-bond from around the superabrasive grain. The question arises now: when is the "sticking" operation complete? Most publications say that when the stick begins to wear rapidly, under only slight hand pressure, the dressing process is complete. Unfortunately, it appears that in practice the stick wears rapidly *all* the time, and proper sticking of a large diameter wheel may require a number of sticks. The grinding wheel periphery will be very smooth to the touch after truing. After sufficient dressing it will feel rough and aggressive, an indication that successful dressing has taken place. "Sticking" is a very arbitrary operation and does not lend itself to CNC or automated grinding operations, particularly when the grinding wheel is wide and narrow sticks are being used. It is important to manually "stick" by first feeding the stick into the center of the wheel and moving slowly across to the left-hand edge, and then feeding into the center once again and moving across to the right. Needless to say, an operator does not particularly enjoy what is a rather dangerous task to perform for wheels 250 mm (10 in) and larger. Semiautomatic "sticking" devices have been manufactured, but they tend to fail in the grinding environment, being particularly prone to the jamming of the sticks in the mechanism (see Fig. 3.9).

Truing and dressing together can be more easily accomplished by machining a mild steel block on a surface grinder or using a mild steel roll on a cylindrical grinding machine with an infeed 0.01 to 0.02 mm (0.0004 to 0.0008 in). The swarf generated when machining mild steel is sufficiently aggressive to erode the resin matrix in front of the grain, as well as loosen and pull eccentric grains from the periphery and true the wheel. This is a particularly good method for truing and dressing, as there is clear evidence that the grinding wheel is ready for use once the chatter marks of eccentricity disappear. The very aggressive

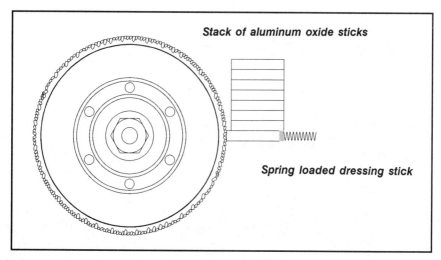

Figure 3.9 Dressing a resin-bonded wheel by sticking.

nature of the steel swarf, however, tends to damage the edge of the wheel by excessive chipping. Softer resin-bonds can also be damaged to a considerable depth by the penetration of the invasive mild steel swarf.

A novel, yet somewhat expensive, system called a "Roll-2-Dresser" both trues and dresses a resin-bonded wheel more accurately than any of the previously mentioned systems. The Roll-2-Dresser (see Fig. 3.10) consists of two rotating steel cylinders, which are driven in the same direction, but at slightly different speeds. The grinding wheel is brought into contact with the cylinders and moved axially across their periphery. The grinding wheel is fed into the rotating rollers with the objective of machining them to generate an aggressive swarf. The axial motion results in a very flat wheel periphery with exceptionally good edge retention.

3.4 Single-Point Dressing

The most common method for dressing a vitrified grinding wheel is to use a mounted diamond point. The mounted point may have only a single diamond or have multi-pointed diamonds. The diamond can be mounted on the table of the grinding machine or be part of a sophisticated system called a pantograph or tracing attachment above the grinding wheelhead (see Fig. 3.11). With CNC control, these complex mechanical devices are unnecessary, as the tracing of the diamond path is easily and more accurately accomplished using the computer control (see Fig. 3.12).

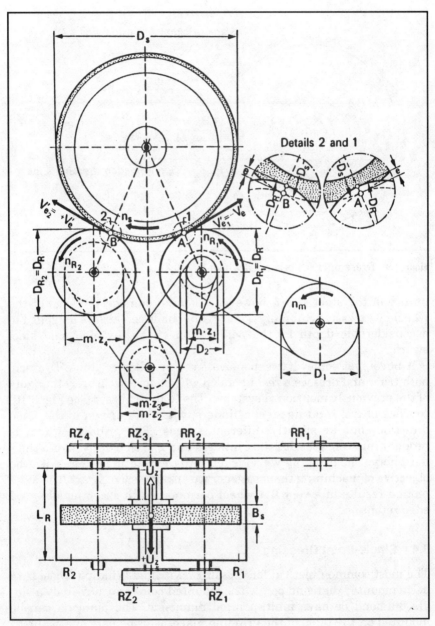

Figure 3.10 Roll-2 dressing system. (*From work by W. Sawluk at Ernst Winter in Germany.*)

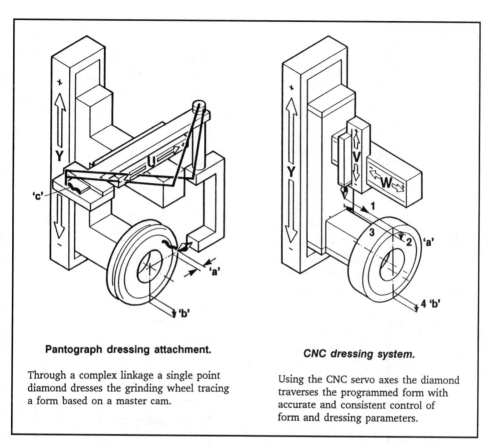

Pantograph dressing attachment.

Through a complex linkage a single point diamond dresses the grinding wheel tracing a form based on a master cam.

CNC dressing system.

Using the CNC servo axes the diamond traverses the programmed form with accurate and consistent control of form and dressing parameters.

Figure 3.11 Pantograph and CNC dressing.

Single-point diamond dressing is carried out using a single-point diamond, set at an angle to the grinding wheel (see Fig. 3.13). A single-point diamond cuts into the periphery of the grinding wheel much like a lathe cutting tool turns a piece of bar-stock and requires clearance angles and attack angles in order to perform its function efficiently. If the diamond were stood upright to dress the grinding wheel, it would wear very quickly. If the set angle of the diamond were equal to the facet angle on the diamond point, it would generate a great deal of heat. The heat would not only destroy the mount, but the thermal growth would upset the accuracy of the traced form as well.

There are many types of mounted point dressers. The single-point is a special case and must be treated with care. The diamond must be rotated during its life to give even wear around the point. The diamond must not be abused, hit, dropped, or allowed to heat excessively, as the point may chip or become distorted in its mount and upset the

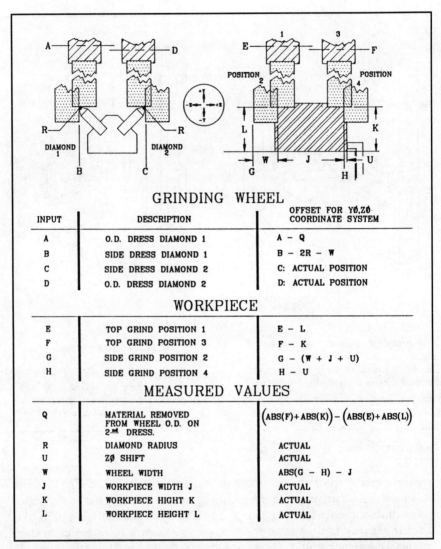

Figure 3.12 Computer controlled dressing. (*From an SME paper by J. A. Gloviak at Bridgeport-Harig.*)

accuracy of the system. Flood cutting fluid must be used at all times when dressing.

Multi-point diamonds and impregnated points have more than one active diamond point in contact with the grinding wheel periphery. They are far more robust and can take a certain amount of abuse. Multi-point systems are often used in very rough grinding operations where precision is not of major importance. Multi-point systems are

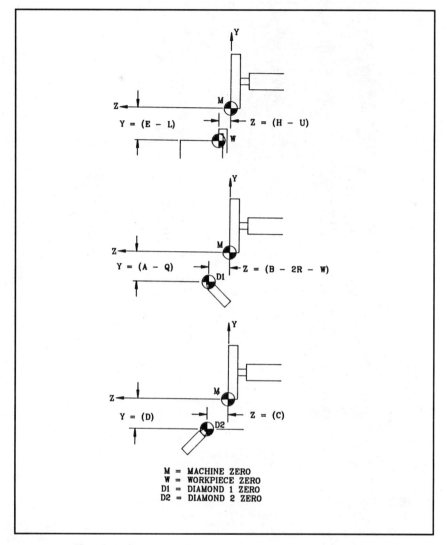

Figure 3.12 *(Continued)*

less accurate than single-points and cannot achieve the fine detail of precise and delicate forms.

Another type of diamond dressing tool is available called a "fliese" tool (see Fig. 3.14), which has shown extended life and better performance over a single-point diamond. "Fliese" is a German word for tile, the name given to this type of tool by Ernst Winter and Company, who claim to have invented it. There are many tools of this type on the mar-

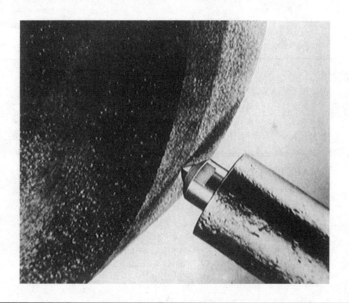

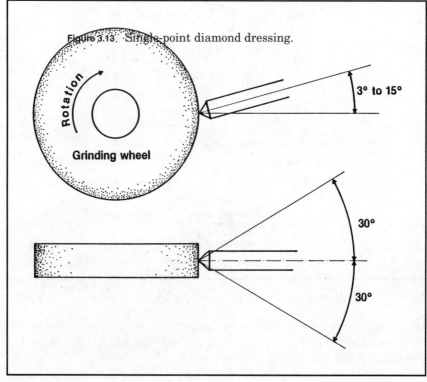

Figure 3.13 Single-point diamond dressing.

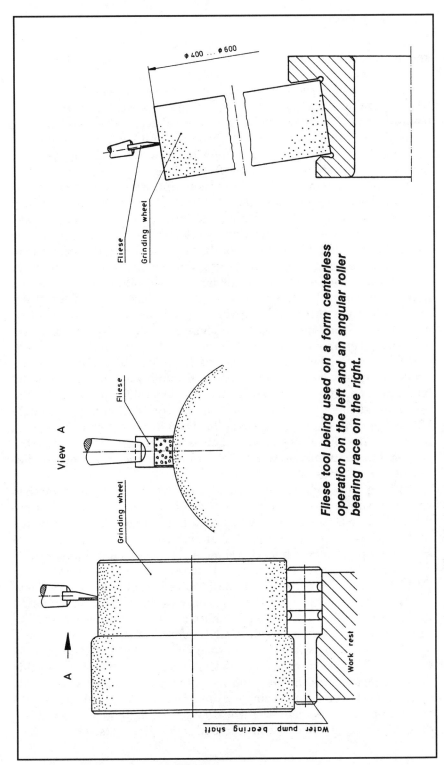

Figure 3.14 Fliese tool dressing. (Courtesy of Diamond Winter Inc.)

ket made by other manufacturers, e.g., "Wega" tools by the Wendt Corporation.

These diamond point dressers are traversed across the grinding wheel, forming the profile and sharpening the wheel periphery. A fast feed will result in a coarse helix to be dressed across the wheel, leaving a periphery suited only for rough grinding, and yielding poor surface finishes. A slow feed results in a finer helix, which, although yielding high surface finish, tends to dress flats on the abrasive grain and prevent high stock removal. It is important that the dressing process is controlled carefully for consistency. It would appear that the industry uses fast dressing and slow dressing to create rough or finely dressed wheels respectively. The question to ask is what is fast and what is slow? The diamond has a finite width, so a calculation can be made to determine diamond overlap. If the grinding wheel is 400 mm (16 in) in diameter and rotates at 30 ms^{-1} (6000 sfm), the RPM will be approximately 1400 rpm. If the diamond is 0.25 mm (0.010 in) wide at dressing depth, then the fastest traverse speed for the diamond not to overlap is 355 mm/min (14 in/min). There is a rule of thumb which recommends overlaps: for roughing there should be two to three overlaps and for fine finishing four to six overlaps. The traverse feed rates would be approximately 150 mm/min (6 in/min) for roughing and 60 mm/min (2.5 in/min) for finishing. Wear will be taking place on the diamond and the effective width should be measured after approximately 325 cm^3 (50 in^3) of grinding wheel has been dressed (see Fig. 3.15). A benefit of the fliese tool is that it does not grow wider as it wears deeper, but maintains a width determined by the width of the diamonds plated to the tile.

Profiling with a diamond point is a relatively inexpensive method used to create a form on a grinding wheel, however, the speed, accuracy, and consistency of the wheel conditioning does not quite match that of diamond roll dressing. A diamond roll is very expensive and sometimes prohibitive for the job shop. A CNC-controlled diamond disk or wafer roll dressing system may be more economical, depending on the detail and the accuracy of the required form. Due to the finite radius on the periphery of the disk, only profiles of moderate detail may be achieved by this method. A diamond disk or wafer roll is basically a very thin diamond roller. CNC control of the thin disk's path across the grinding wheel face allows a form to be dressed quite inexpensively. The dressing parameters that need to be controlled with this dressing method are the diamond wafer to grinding wheel speed ratio, and the traverse rate, as in single-point dressing. The resulting peripheral condition of the grinding wheel is almost as though it had been dressed with a full form width diamond roller dresser (see Fig. 3.16).

Abrasive Preparation 63

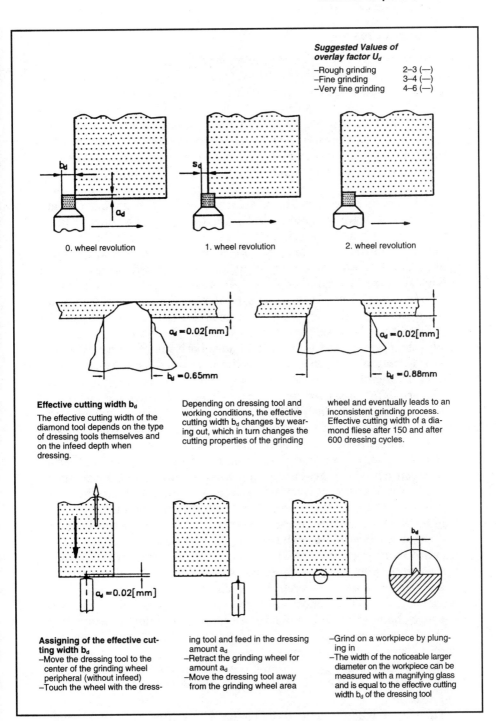

Figure 3.15 Principles of single-point dressing by overlaps. (*Courtesy of Studer AG.*)

3.5 Crush Dressing

A natural progression from mounted point dressing of complex profiles came about because of the time involved in carefully tracing a single-point diamond across the wheel periphery. Mass production demanded faster dressing times. Crush dressing was adopted. Crush dressing is a dressing method which uses a hard roller made from hardened steel, high-speed steel, tungsten carbide, or boron carbide to dress a vitrified bonded or crush dressable, metal-bonded grinding wheel. The material choice for the crush roll is made with two features in mind: wear resistance and stiffness. The crushing roll has the profile of the part to be ground machined around its periphery. The hard crushing roll and grinding wheel are brought radially together, and the roller literally crushes its way into the grinding wheel, breaking the grain from the bond as it forces and impresses its shape into the grinding wheel periphery. Crushing is carried out at low wheel speed, typically 0.5 to 1.5 ms^{-1} (100 to 300 sfm). The crush roll can be motor driven to drive the grinding wheel. Conversely, the crush roll can be motor driven as it contacts a rotating grinding wheel and once the peripheral speeds have equalized the crush roll motor is disengaged. The grinding wheel then drives the dresser. It is important to have both the grinding wheel and the crush roll rotating on contact to eliminate the possibility of grinding a flat on the crush roll, thus imparting a vibration into the system. The radial crushing forces are very high (see Fig. 3.17). Such high forces are the cause of process instability, which can prevent an intricate form from being dressed or cause deflection in the system, which imparts lobes on the wheel and very poor surface finish on the workpiece.

Research has shown that for a given machine tool system and crush roll in contact with a given grinding wheel, rotating at a given speed, there is a stable width of contact between the crush roll and the grinding wheel. If that width is exceeded, the result will be unstable vibration between the two rotating members. Factors which will influence the onset of vibration are the rigidity of the bearing system and the balance of the rotating members. Crush dressing is most prone to vibration because of the high forces between the roll and the grinding wheel, as well as the generally weak design of dressing systems on machine tools. Diamond roll dressing is less prone to vibration because the area of contact for the same form width of dresser is less and the speed ratios are much higher. The requirement for extreme rigidity when crush dressing is emphasized by many machine tool builders, who make provision for outer supports, on both the cantilever grinding spindle and dresser unit, specifically for crush dressing applications.

Take care when crush dressing on a machine with a precision

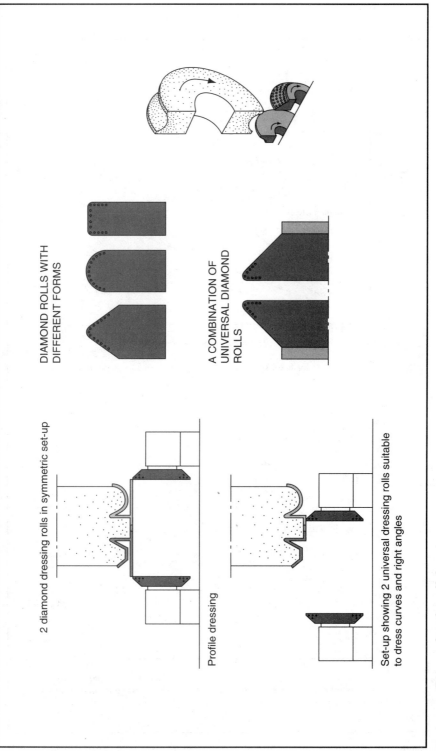

Figure 3.16 Diamond wafer roll dressing.

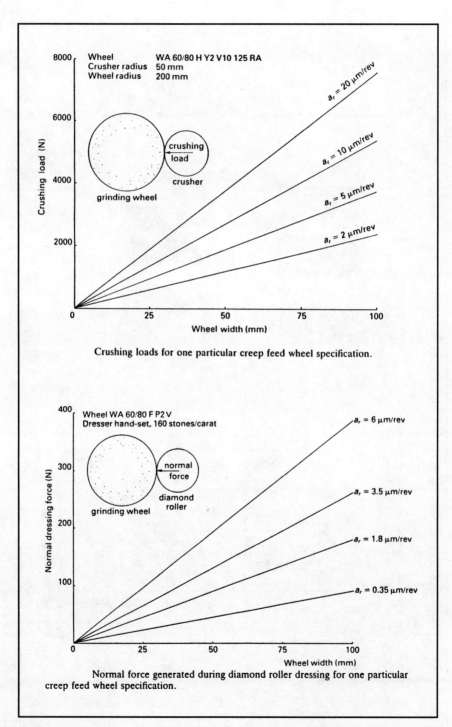

Figure 3.17 Comparison of forces when crush dressing and diamond roll dressing.

rolling, contact ball bearing spindle. Prolonged use of high crush dressing forces on a machine of this construction can cause heavy wear on the grinding spindle bearings and loss of spindle rigidity, leading to poor surface finishes and loss of accuracy. Heavy-duty spindle designs using either roller bearings, with a larger area of contact, or hydrostatic bearings are better suited for the crush dressing method.

The crushing unit is usually table or wheelhead mounted. Heavy wear takes place on a crush roll, so in high production two rolls are usually positioned at either end of the grinding table, so that one is used as a roughing roll and the other used only for finishing and lighter dressing. Of course, the grinding wheel can be used to refurbish a worn roughing roll, making crush roll dressing quite economical. Medium grade, 120 to 400 grain grinding wheels are typically crush dressed. Constant pressure should be kept between the wheel and roll during the crushing. It is therefore essential to maintain an infeed proportional to the number of revolutions of the grinding wheel, typically 0.002 to 0.02 mm (0.0001 in to 0.001 in) per revolution of the grinding wheel. High-quality cutting fluid should be applied during crushing. Heavy oil is preferred due to the presence of high-pressure (hp) additives.

Crush roll dressing will generally yield the sharpest grinding wheel. Depending on the structure of the grinding wheel, crush dressing can actually compress and push debris into the porosity of the wheel and make it unsuitable for grinding. It is important to use a high-pressure jet of cutting fluid directed toward the wheel periphery to blast away any free grains which might otherwise become trapped in the wheel and then released in the grinding operation, upsetting the surface finish with occasional slash marks in the surface being ground. Due to the wear characteristics, only moderate detail of form may be successfully and economically retained. Crush dressing is relatively inexpensive, yet may be time consuming due to the need to stop the grinding process, slow the grinding wheel down, and slowly feed the crush roller (perhaps even two rollers: one a roughing and one a finishing roller) into the grinding wheel prior to machining the next part. Though a boron carbide roll has the best wear resistance, the material is very brittle and can vibrate at very high frequencies, therefore preventing deep and delicate forms from being dressed into a grinding wheel before reaching full form depth. A compromise in wear resistance will have to made in this case, choosing high-speed steel over the boron or tungsten carbide, for a more stable system.

3.6 Diamond Roll Dressing

For most abrasive machining processes the grinding wheel has to be conditioned in some manner to allow the grinding wheel to perform with a high level of efficiency. Some dressed forms are simply plain forms. Plain surface or cylindrical roll grinding is where a flat plain surface is machined. A flat, plain form is not generally regarded as a form profile. Dressing such a plain conventional grinding wheel is typically carried out using a mounted diamond point. The mounted point is passed across the face of the grinding wheel periphery. The traverse speed of the diamond point will determine the condition of the grinding wheel. The traverse speed of the diamond point is the controlling factor in producing a sharp or dull grinding wheel. Producing complex forms by tracing with a single-point diamond leaves little room for fine detail and accuracy. This is due to the difficulties in varying the tracing speed and angle of attack across the form, and, at the same time, determining and compensating for diamond wear.

Modern-day, high-production grinding machines have the ability not only to machine to high degrees of surface finish and size, but also to achieve the most accurate forms in a variety of the most difficult to machine materials. It is in the development of such systems, recognizing the shortcomings of crush roll and single-point dressing, that diamond roll dressing has become the only process available to consistently and economically produce the desired results.

There are five main reasons to choose diamond roll dressing (see Fig. 3.18):

1. Formed profiles

2. Consistent accuracy of size and form

3. Control of grinding wheel sharpness

4. Control of workpiece surface finish

5. Economy of production

The demands put upon modern-day engineering components require greater attention to form profiles. The shapes of many workpiece surfaces are complex and have extra-fine detail within extremely close form tolerances. The requirements of accuracy are ever increasing as part performance demands tighter tolerances. High accuracy has to be achieved in both everyday, easy-to-machine materials and superalloys and more difficult-to-machine materials. The accuracy of the finished part depends largely upon the accuracy of the profile shape, concentricity, and runout of the diamond roll. However, without an equally rigid and accurate dressing unit (see Fig. 3.19), the system breaks down.

A diamond roll is a most versatile dressing tool. It can achieve a grinding wheel sharpness which virtually equals that of crush dressing, yet, at the same time, may achieve more accurate form and more

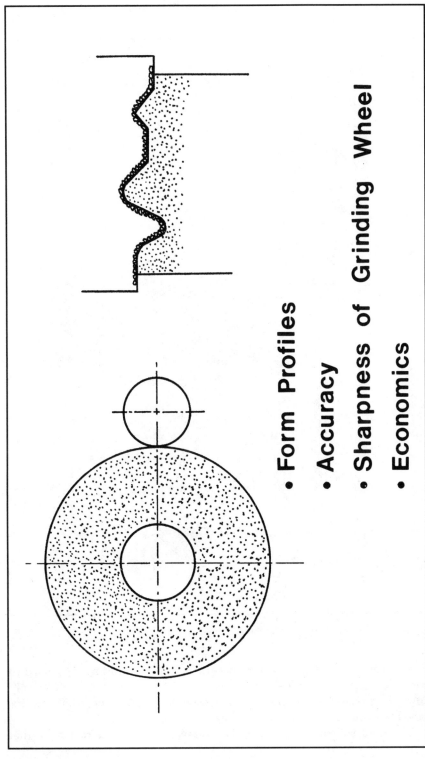

Figure 3.18 Diamond roll dressing.

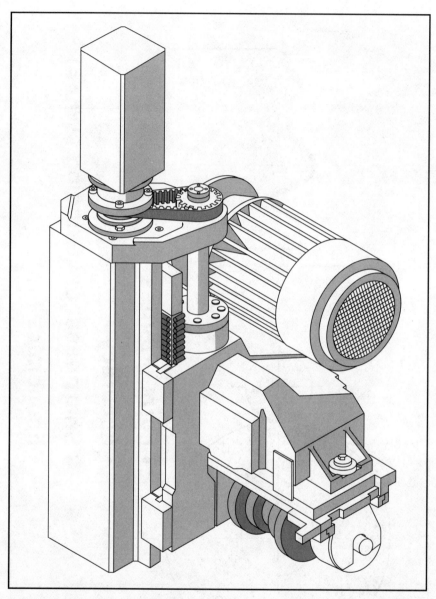

Figure 3.19 Diamond roll dressing unit.

closely controlled dressing parameters. A diamond roll may be used to condition a grinding wheel in many different modes which effectively tune the process to the needs of the economic manufacture of the finished workpiece.

Diamond roll dressing not only achieves a full form width profile, dressed at wheel speed with an accuracy not achievable by any other technique, but also achieves a high, consistent, long-lasting standard. For many applications diamond roll dressing is the only viable method. Where other systems of dressing are viable, diamond roll dressing may still be cost-effective, depending on batch quantity, the cost of each individual workpiece, and the flexibility required in the machining system. The high initial cost generally rules out the use of a diamond roll for one-off job shop applications. However, as the batch quantities become larger, diamond roll dressing becomes the more economical choice. There are three main types of diamond roller dressers (see Fig. 3.20):

1. *Hand set, sintered.* This is a diamond roller which has had the diamonds hand set around a female mold of the form and then sintered into a solid roll by the addition of a powdered metal between the roll core and outer shell (see Fig. 3.20). These rolls are the least accurate due to the thermal and mechanical distortions which take place during the sintering process, necessitating subsequent lapping of the diamonds on the roll and resulting, in essence, in a worn roll at the beginning.

2. *Hand set, reverse plated.* This type of roller is a very robust roller and best suited to relatively coarse forms of large radii and open tolerances. The diamonds are hand set with glue on the inside of the female form of a graphite mold. A nickel core is placed in the center of the mold in a plating bath, where the nickel is allowed, over a period of time, to build up and form a shell of accurately positioned diamonds imbedded in a nickel matrix. Once a sufficient depth of nickel has formed, then a stainless steel core is positioned in the center of the shell and the two are cemented together with a low melting point brass type alloy. Depending on the quality of the initial female mold and the plating process, some minimal lapping of the diamonds may be necessary.

3. *Random set, reverse plated.* The process is exactly the same as the hand set, reverse plated roll, except that the diamonds are not glued to the female form. The diamonds, usually much smaller than those used for handsetting, are electrostatically oriented or held in position by a centrifugal force. These diamond rolls are the most accurate, achieve the finest detail, and usually have the longest life because of the large diamond coverage. As there is no lapping involved in the manufacture of these rolls, they are generally the most aggressive and can dress a very sharp, cool cutting periphery on the grinding wheel. Industries which require the fine form detail, yet do not require the aggressive grinding, specify the roll to be

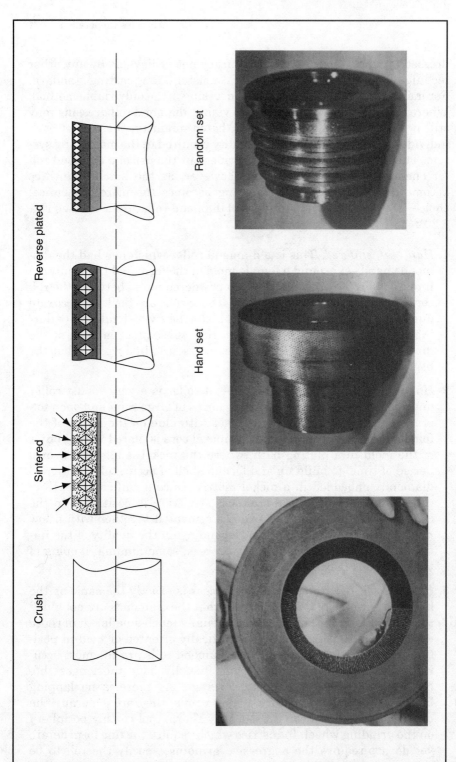

Figure 3.20 Types and construction of diamond rollers.

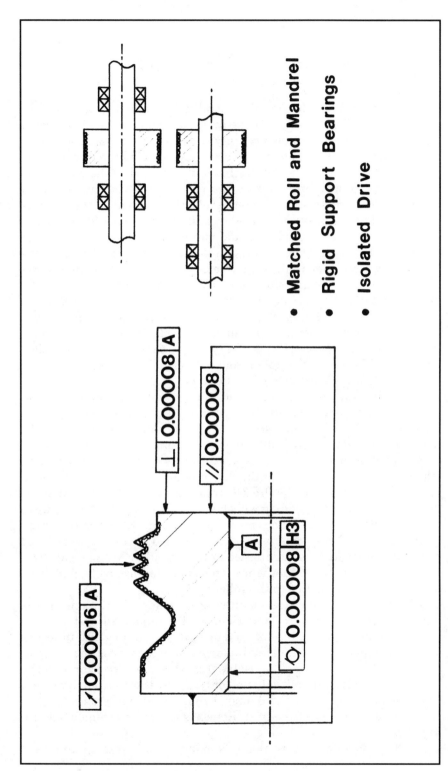

Figure 3.21 Accuracy of diamond roller dressers.

lapped in order to achieve the required surface finish. This is particularly the case for the bearing industry.

It is important to note that a diamond roll is made to exacting tolerances (see Fig. 3.21) and unless the dressing unit onto which the roller is mounted is of comparable accuracy then there is no hope of achieving the high standards of form and accuracy expected of a diamond roll dressing system. The diamond roll is best matched to the mandrel of the roll unit in order to minimize run out between the roll periphery and the bearing diameters. When the roller is mounted on the mandrel and located in the dressing unit, the radial run out should not exceed 0.002 mm (0.00008 in) and the axial run out should not exceed 0.001 mm (0.00004 in) TIR. The dressing unit must be rigid and support the mandrel in a manner that minimizes deflection, whirling, and vibration. The drive to the dressing roll must be isolated both mechanically and thermally. Dresser motor drives have to run at relatively high speeds, opening up choices to air motors, hydraulic motors, and electric motors. The choice of drive motor is critical. Air motors are most unsuitable, as they lack the necessary torque to maintain a controlled speed ratio between the roller and the grinding wheel. Hydraulic motors generate a great deal of heat and may seriously affect the accuracy of axial run out of the diamond roll, as well as its axial positioning. The heat generated in the hydraulic system, without suitable cooling, will decrease the viscosity of the oil and therefore cause the motor to slow down. Hydraulic drives also present a problem when running a dressing roll at close to synchronous speed with the grinding wheel. At a ratio of 0.8 to 0.9 the dressing unit has to act more like a brake in order to avoid the dresser rotating at wheel speed. At these high-speed ratios the grinding wheel will tend to drag the dresser up to a synchronous speed which must be avoided. Electrically driven motors of 2 kW (2.5 hp) minimum should be used in order to be assured that the dresser RPM is maintained at the set value. Mechanical isolation is also necessary so that vibration from couplings, belts, and drive motors is not transmitted to the roller.

A diamond roll should be mounted on a mandrel to maintain the highest precision, balance, and rigidity. The preferred system is to mount the roller on a plain shaft and use a highly torqued nut to secure the roller in place. It is often felt by some that a keyway is necessary to drive the roller. This is not advisable. Keyways machined through the bore of the diamond roll will not only upset the concentricity and weaken the structure of the diamond roll, but also affect the overall accuracy and balance of the unit. Keyways are not recommended (see Fig. 3.22).

Fig. 3.23 shows a cross section of a diamond roll dressing unit built

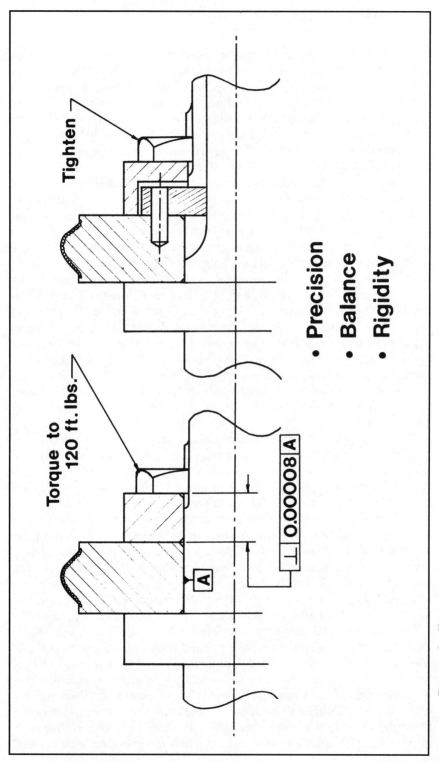

Figure 3.22 Diamond roll mounting.

for research purposes. A number of lessons may be learned from this design. First, rigidity: the mandrel diameter is 52 mm (2 in) and the double row angular contact bearings are spaced as close as possible to one another, either side of the roller, for maximum rigidity. This system, as with most production dressing units, has a minimum rigidity of 0.3 GNm^{-1} (1,700,000 lbf/in). The unit is also dynamically balanced for high speed—up to 5,000 rpm. The seals shown are a contacting type of neoprene seal which are unsuitable for this application in production. Contacting seals generate a great deal of frictional heat and produce thermal errors. Neoprene also reacts with the additives in most premium cutting fluids. It is recommended that high-speed dressing systems have positive pressure air/oil mix lubrication to insure positive pressure inside the bearing housing and prevent the ingress of cutting fluid and grinding detritus. The whole dressing unit should be isolated from its drive by an isolating coupling. The infeed mechanism of the dressing unit should also be very stiff, in order to resolve very small incremental infeeds and infeed rates per revolution of the grinding wheel for continuous dressing. The dresser drives should also provide rapid exit capability from the grinding wheel after an intermittent dress cycle in order to minimize the dresser dwell time. A unit called a "swing-step" dresser, available from the KW company in Germany, minimizes the dwell time using a rotary dresser fixed to a rigid bearing. The dresser is positioned so that, as it swings past the grinding wheel, it is in contact for a minimal time (see Fig. 3.24).

The need for grinding wheel sharpness is most important to the success of the grinding process. Research has shown that most of the grinding energy is consumed in rubbing energy. This is energy where the grinding wheel becomes dull and rubs the surface of the workpiece instead of cutting cleanly. This makes the control of the sharpness of the grinding wheel, with respect to the growth of wear-flats, most important. Poor dressing technique can actually dress flats onto the periphery of the grinding wheel, causing burning and very bad grinding conditions. The diamond roll may have dressed a perfect profile on the grinding wheel, yet the sharpness of the grinding wheel periphery has been dulled by the action of the diamonds cutting through the abrasive grain and effectively making the cutting edges flat and dull.

The diamond roll dressing technique has the ability to very accurately control the grinding wheel performance and therefore the process. The ratio of rotation of the diamond roller to the grinding wheel is important since the diamond roller can condition the grinding wheel to be very sharp and open or quite dull and closed. As the grinding wheel dulls, the surface finish on the workpiece improves because the grinding wheel is performing more of a burnishing action rather than keen cutting. It must be realized that when the grinding wheel is dull

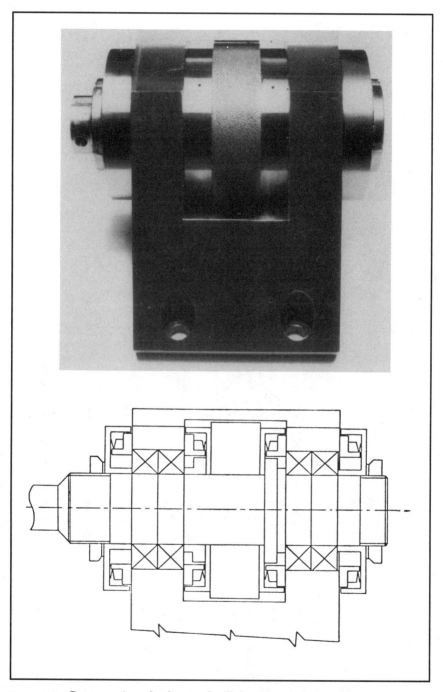

Figure 3.23 Cross section of a diamond roll dressing unit.

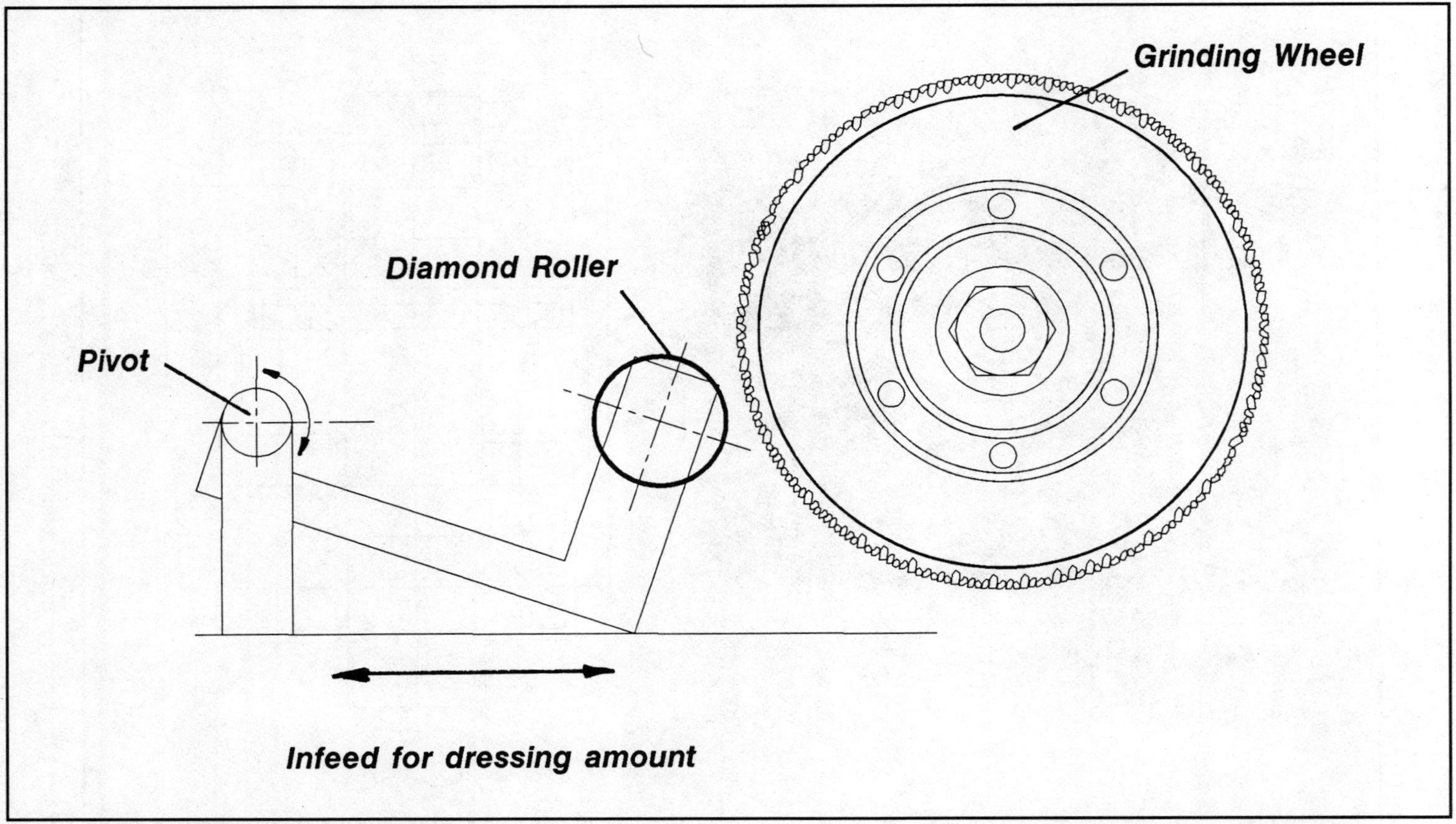

Figure 3.24 Swing-step dresser. (*Unit by KW.*)

the surface finish might be good, but the heat input to the surface of the workpiece may be detrimental and cause metallurgical damage.

Research has found there to be a relationship between the speed ratio of the diamond roll and the specific energy of the grinding process. It is important to keep the specific energy as low as possible in order to protect the workpiece from thermal damage. The lowest specific energy occurs at a ratio of 1:1 (as in crush dressing). Unfortunately when the dresser and grinding wheel run synchronously, as in crush dressing, the diamonds and the abrasive grain interlink and the diamonds are quickly plucked from the dresser, causing expensive and heavy wear. A ratio of 0.6 to 0.9 yields the ideal conditions for most high-stock removal applications, in particular creep-feed grinding. It should be appreciated that the ratio between the diamond roll and the grinding wheel may be used to tune the process. A ratio as low as 0.2 to 0.4, or even, with the roller rotating in the opposing direction, as low as 0.2 to 0.4, will yield a much finer surface finish, as the roller has dressed flats on the surface of the grinding wheel (see Fig. 3.25). Though a finer surface finish results, the specific energy will be high and the risk of thermal damage will be much greater. It should also be recognized that as the surface speed differential between the grinding wheel periphery and the diamond roll increases, the life of the diamond roll decreases. The specific energy may also be affected by the radial infeed rate of the dressing roll.

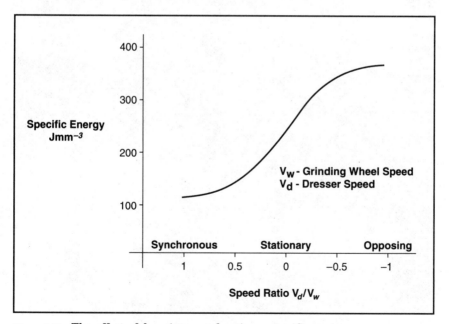

Figure 3.25 The effect of dressing speed ratio on specific energy.

Different dressing methods and ratios also cause different forces on the dressing system, which emphasizes the need for high rigidity and vibrational stability. Diamond roll dressing creates less dressing force compared with crush dressing and is generally carried out at full wheel speed. The life of a diamond roll is very long, considerably longer than a crush roll, even though some of the most accurate and fine forms may be held using this method—typically as fine as 0.12 mm (0.005 in) radius and less than 0.005 mm (0.0002 in) tolerance on form.

Figure 3.26 Overhead dressing system.

Diamond rolls may be table mounted or overhead mounted (see Fig. 3.26). Typically, high production machines use overhead dressing systems for fast dressing cycles, as well as to facilitate the opportunity to use continuous dressing. It is important, in intermittent dressing, to minimize the dwell period of the dresser on the grinding wheel. Excessive dwell time, due to a poorly maintained machine, can lead to a very dull wheel in a situation where the wheel has apparently been dressed aggressively (see Fig. 3.27).

Particular care must be taken when mounting and adjusting diamond rolls. Rolls fit on the dresser shaft to very close tolerances +0.008 −0.000 mm (+0.0003 in −0.0000 in) on a 52 mm (2 in) diameter and require care, patience, and cleanliness in setup. The diamond roll should be clamped firmly and evenly in the dresser unit to avoid distortion—packing material and/or shimming in order to adjust the dresser position with washers is inadvisable; proper spacers have to be manufactured within the same tolerances that the diamond roll was

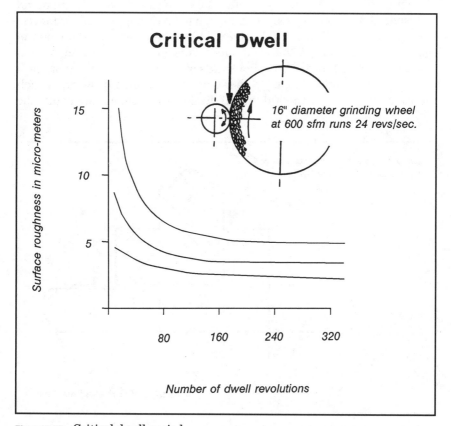

Figure 3.27 Critical dwell period.

produced. To assist in the removal of the diamond roll, two thin, soft, steel sacrificial shims can be made to fit between the flange face of the dresser mandrel and the mating face of the diamond roll. To remove the diamond roll, the tendency for the two mating faces to stick can be broken by driving a screwdriver between the shims. The dresser and the flange face are left undamaged, and the diamond roll is now floating on the mandrel for easy removal (see Fig. 3.28). The mounted dresser should run true and in-line with the grinding spindle. It is most important to keep a copious flow of cutting fluid between the diamond roller and the grinding wheel at all times. The diamond roll must be kept cool to avoid damage, as well as to maintain thermal and, therefore, dimensional stability. Second, the free grain which is present during dressing must be washed away from the dressing zone. The free grain must be prevented from coming between the grinding wheel and dressing roll and effectively eroding away the diamond roller matrix.

A diamond roll deteriorates and performs unsatisfactorily when the diamonds fall out of the matrix. Fig. 3.29 shows six stages of diamond roll wear: A shows the dresser as new, with a strong backup of matrix surrounding the diamonds. B and C show the progressive erosion of the matrix in front of the diamonds. This progression from A to C takes place over hundreds of thousands of dress cycles. It is at stage D where there might begin to be traces of diamond wear, which increases the load on the individual diamonds. Also, at this point, the

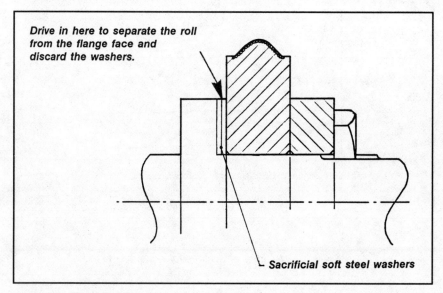

Figure 3.28 Diamond roll removal.

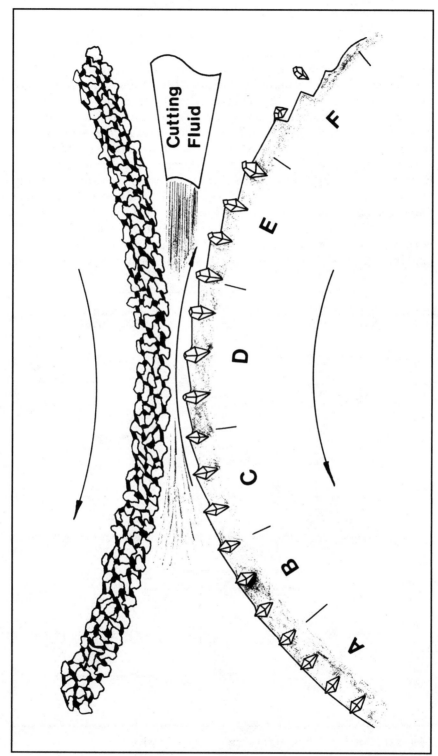

Figure 3.29 Diamond roll wear.

form is still good on the dresser, but some diamonds may have become dislodged from the matrix and a hole remains. It is at this time that the dresser can be "repaired" by the roll manufacturer by positive plating either nickel or chrome onto the roll and filling the voids left by the missing diamonds, as well as reinforcing the worn matrix around the existing diamonds. This may be done only once or twice in the life of a diamond roller. At E, the bonding matrix becomes very weak, the diamonds are worn, and the force on the diamonds is high. Eventually at F, the diamonds fall out, the form is lost, and a new diamond roller is required to continue operations. However, it is seldom found that a diamond roll fails because of this. In the shop environment, it is unfortunately prevalent for the diamond roll to be abused with hammers or mallets, maltreated by being run dry, or fed into a grinding wheel without the dresser drive being switched on. Seldom is

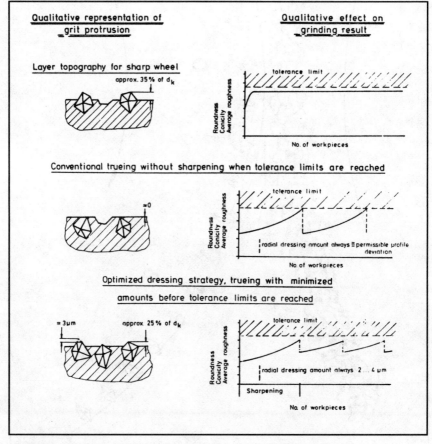

Figure 3.30 Optimized dressing system for superabrasive wheels.

the life of the diamond roll enjoyed to the maximum because of handling malpractices.

The advent of the newer microcrystalline CBN abrasive grain demands extremely fine resolution of the dresser infeed mechanism. Microcrystalline CBN abrasive allows a fine truing operation to be carried out without the need to dress and open the wheel face. In effect, a slight loss of form tolerance may be replenished by an extra fine truing of the wheel periphery, which does not penetrate into the wheel matrix, and therefore requires no dressing. The economics are good, both in abrasive saving as well as dress time saving (see Fig. 3.30).

3.7 Continuous Dressing

In all of the discussion on grinding wheel preparation, the grinding wheel was trued and dressed prior to grinding, so the properties of the abrasive grain, bond system, and wheel structure dictated the life of the wheel before truing and dressing had to be carried out again to refurbish the grinding wheel surface for efficient grinding. The grinding wheel was at its peak sharpness and maximum potential performance just after dressing. As soon as the grinding wheel touched the workpiece material it had degraded and the process began to gradually fall off in efficiency. Out of a comprehensive creep-feed grinding research program, the continuous dressing process was developed. Using a diamond roller in constant contact with the grinding wheel, phenomenal stock removal rates can be achieved, even in the most difficult to machine materials. Although the research was carried out using a mechanical system of linkages, continuous dressing is best carried out using a CNC grinding machine. The dressing process is controlled by the differential in speed between the rotating diamond roller and the grinding wheel and the radial infeed rate of the diamond roller into the grinding wheel (see Fig. 3.31).

3.8 EDM Dressing Metal Bond Systems

Metal bond wheels have been manufactured in crush dressable matrices. However, the great majority are sintered in copper/bronze materials and even cast iron. Truing and dressing these types of wheels to accurate forms and with a periphery open enough to cut is most effectively done using Electro-discharge Machining (EDM). The electric arc readily erodes the matrix as well as the abrasive, and, depending on the current density, the wheel surface can be prepared with coarse pits, which allow chip clearance and cutting fluid application. EDM dressing of metal-bonded wheels is usually performed off the grinding machine and on an EDM machine. There has been a ceramic grinding

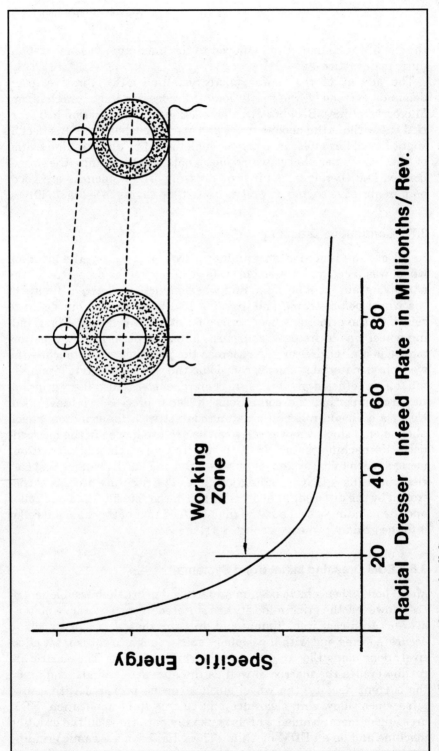

Figure 3.31 Continuous diamond roll dressing.

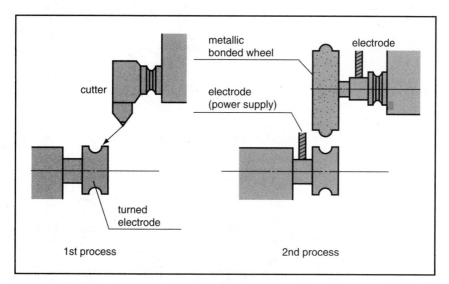

Figure 3.32 Electro-discharge Machine (EDM) dressing. *(Courtesy of Mazak Corporation.)*

machine produced which combines the whole EDM process in the grinding machine (see Fig. 3.32). The CNC control forms a carbon electrode, and, using the cutting fluid as the electrolyte, EDMs the profile in the wheel. Over a period of time, however, the electric current will cause detrimental pitting of the spindle bearing. It is therefore better to perform EDM dressing off the grinding machine.

Chapter 4

Fundamentals of Grinding

4.1 A Micro-milling Analogy

The grinding process may be described as a machining process which is analogous to the milling process, using a milling cutter having thousands of very small teeth. Each tooth removes an extremely small chip from the surface of the workpiece material, and produces a smooth and accurate surface finish. In our analogy, the cutting teeth will be the abrasive grains, which are all different and irregular in shape and randomly oriented in the grinding wheel. The grinding process is very difficult to analyze and even more difficult to model due to its stochastic nature. An attempt to model the process will, however, provide us with a better overall understanding and help us to interpret the effects and the interactions of the process; that is, if our conclusions are made intelligently and within the bounds and limitations of the model.

The intention here is to develop a piece-by-piece relationship among all the constituent parts and aspects of the abrasive machining process into a mathematical model and formulate a generic expression for the process.

A mathematical relationship to the grinding process is a worthwhile approach to develop an understanding of the process; moreover, it allows us to perform "what if" scenarios. What happens if the wheel speed is increased? What is the effect of taking a larger depth of cut or a faster feed rate? Knowing the geometric and mechanical relationships of a mathematical model can guide us toward an optimum in grinding performance, provided our model is a good one. The following micro-milling analogy is a well-tried model of the process, but notice as the model develops, how there are certain assumptions made along the way. These assumptions will bound the model, so it is important to remember those limits in the final analysis. The micro-milling analogy is at best a guide and does not represent the exact mechanism by which the grinding process operates.

Figure 4.1 illustrates the geometric relationship of our micro-milling analogy. The following nomenclature will be used in the mathematical equations:

Grinding wheel diameter	D (mm)
Peripheral speed of the grinding wheel	V_s (ms^{-1})
Angular speed of the grinding wheel	ω (rad/s)
Workpiece infeed rate	V_w (mms^{-1})
Number of grains along a peripheral line	K
Number of active grains per unit area	C

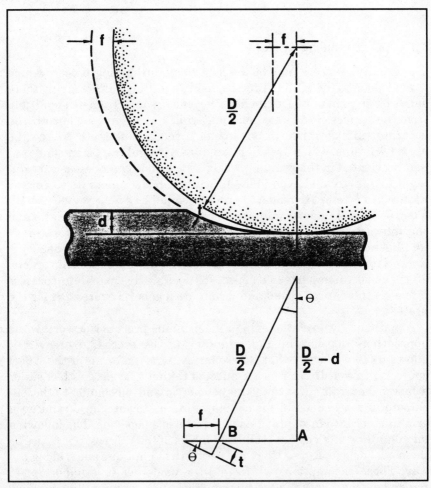

Figure 4.1 Geometric diagram of the micro-milling analogy.

Grain depth of cut	t (mm)
Length of the arc of cut	l (mm)
Grinding wheel depth of cut	d (mm)
Theoretical width of an individual grain	b' (mm)
Grinding wheel width of cut	b (mm)

In Fig. 4.1, the grinding wheel depth of cut is greatly exaggerated (though perhaps not for creep-feed grinding). The length of the arc of cut, l, may be approximated as follows:

$$l^2 = D^2/4 - (D/2 - d)^2 + d^2$$
$$= D\,d$$
$$l = (Dd)^{0.5} \tag{1}$$

Refer to the geometric diagram of the micro-milling analogy in Fig. 4.1.

$$AB = ((D/2)^2 - (D/2 - d)^2)^{0.5}$$
$$= (d(D - d))^{0.5}$$

Assume that the arc of cut is a straight line and f is the distance that the workpiece advances in $1/K$ revolutions of a grinding wheel having K grains along a peripheral line.

$$t = f \sin \theta$$
$$= f(AB)/(D/2)$$
$$= (2f(d(D - d))^{0.5})/D$$

Because d is significantly smaller than D, then:

$$t = 2f(d/D)^{0.5}$$
$$f = V_w/(K\omega)$$
$$t = (2V_w/K\omega)(d/D)^{0.5} \tag{2}$$

It is worth commenting on and understanding the above assumptions:

Assumption #1. The arc of cut being a straight line.
The arc of cut is *not* a straight line. However, in the limit of the geometry, the assumption appears valid. For example, the error in the arc length as a straight line over the calculated arc length for a grinding wheel 400 mm (16 in) in diameter and a depth of cut of 0.02 mm (0.001 in) is little more than 1 in 100,000. The fact of the matter is that the grinding forces are high in the area of the arc of cut; high

enough to deform the grinding wheel like an automobile tire deforms on the road. Depending on the wheel strength and the strength of the material being ground, the true contact area can be substantially longer than that calculated geometrically by a factor of 1.5 to 2 times. The only exception in this scenario is creep-feed grinding, where the arc length of cut is very long and the wheel deformation is insignificant with respect to the geometrically calculated arc length.

Back to the analogy: C, the number of active grains per unit area, has to be determined by a suitable method. Then the surface area of the grinding wheel of grains b' wide is $\pi Db'$.

$$K = \pi Db'C \tag{3}$$

Assumption #2. The number of grains along a peripheral line.
It is very difficult to count the number of grains along a peripheral line of a grinding wheel. One method used to count the number of grains on a grinding wheel periphery is to take the dressed wheel and roll it over a piece of flat glass which has been smoked with soot from a wax candle. The grains which protrude from the periphery leave a mark in the soot and can be counted. A more sophisticated technique is to make a plastic replica of a section of the grinding wheel surface and, in a Scanning Electron Microscope (SEM), plot a topographical map of the grinding wheel surface and extrapolate the map to count the protruding grains. One fallacy in these measurements, no matter how sophisticated they may be, is that they are made with the grinding wheel in a relaxed mode. Under the forces of grinding, the bond bridges will deform. In a relaxed condition, what might appear to be inactive subsurface grains will, under the forces of grinding, become active due to the deformation of the grains at the surface. The grains are not equally spaced around the wheel periphery and so there may exist situations where two grains, which would be counted as active, follow directly behind one another, and only one grain is truly active, not two. An exception to the random pattern of the grains will be when "Diamesh" wheel construction is used. All of this estimation also presumes a perfectly concentric grinding wheel with infinite rigidity.

Looking at the frontal area of an idealized grain:

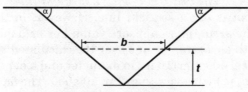

Frontal area of an idealized grain

The ratio of width to depth for plunge grinding $r = b'/t$ for surface grinding and cylindrical grinding. However, the mean grain depth of cut is used $t/2$. Hence $r = 2b'/t$.

Equation (3) now becomes:

$$K = \pi DrtC/2$$

Substituting into equation (2):

$$t = (d/D)^{0.5} (4V_w/\pi DV_s Crt)$$

but $V_s = \pi D\omega$

$$t^2 = (4V_w d/V_s Crl) \qquad (4)$$

Assumption #3. The idealized grain.
Abrasive grains are irregularly shaped. Our attempt at grading the grains by size by riddling them through a wire sieve tells us little about their shape. Truthfully, the wire mesh allows a particle through, which has one plain of cross section the same or smaller than the square cross section of the space between the wires. That means for a gap 0.2032 to 0.2540 mm, which is FEPA B181 (0.0125 to 0.010 in or 80/100 U.S. Mesh), a sphere of that diameter could pass through, as well as a cube, an octahedron, or a long thin needle, all within the confines of the two-dimensional cross-sectional area of the sieve. There is nothing in our grading system which defines the shape or aspect ratio of the grain, so the idealized grain is truly idealized. In fact, it is idealized to the point that there has never been a grain produced, in quantity, to that shape.

Using classic single point cutting geometry:

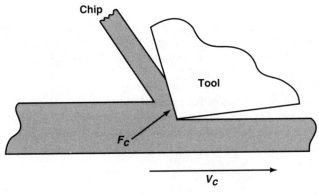

where
- b = width of cut (mm)
- t = undeformed chip thickness (mm)
- u = specific energy (Jmm-3)

For single point cutting $u = F_c V_c / b t V_c$
The cutting force $F_c = ubt$
The cutting force in grinding is attributed to the tangential force, so for grinding, the equation can be modified as follows:

$$u' = F_t V_s / V_w b d$$

The cutting force $F_c = u'bd V_w / V_s$
The knowledge of the mean force acting on a single grain is necessary when considering how the effective hardness of a grinding wheel is altered by the grinding conditions, e.g.:
Work done per unit time is $= u'V_w bd$
Number of working grains per unit time are $= V_s C b$
Work done per grain is $= u'V_w d / V_s C$
The mean force acting on a single grain F'' may be obtained by dividing the work done per grain by the length of the arc of cut:

$$F'' = u'V_w d / V_s C l \tag{5}$$

from equation 4

$$t^2 = 4V_w d / V_s C r l$$

$$1 = 4V_w d / V_s C r t^2$$

substituting equation 1 in equation 5

$$F'' = u't^2 r / 4 \tag{6}$$

By using the model it is possible to trouble-shoot a grinding problem and be given alternative solutions for a course of action which will alleviate the problem.

For example, if a grinding operation were suffering under what appeared to be a grinding wheel grade which was too hard, what measures could be taken to make the wheel act softer?

Based on equations 4 and 5:

1. Increase the workpiece infeed rate – V_w
2. Increase the grinding wheel depth of cut – d
3. Decrease the grinding wheel speed – V_s
4. Decrease the number of active grains in the grinding wheel – C

4.2 Energy Used for Grinding

Already the term "specific energy" has been referred to in the micromilling analogy. Specific energy is the amount of energy which is used to remove a unit volume of material. The units of specific energy are Joules per cubic millimeter, Jmm^{-3} (Btu/in^3). The energy used in grinding is dispersed in seven main ways.

1. Heating the workpiece material
2. Heating the grinding wheel
3. Heating the chips
4. Kinetic energy in the chips
5. Radiation to the surroundings
6. Energy to create a new surface
7. Residual stresses in the lattice of the ground surface and the chip

The most important energy concern to the manufacturing engineer is the amount of energy which is transferred into the surface of the workpiece. High specific energy causes high heat input, high residual stresses, and poor metallurgical integrity.

The lower the specific energy of the process, the less chance there is of damaging the workpiece. It is therefore our objective to minimize the amount of energy used in grinding a given workpiece to ensure that the maximum energy is channelled into the chips and the surroundings, and away from the workpiece surface.

All grinding processes are composed of varying amounts of cutting, plowing, and rubbing energies, depending upon the sharpness of the abrasive (see Fig. 4.2). The cutting energy is that energy which causes shear to occur ahead of the abrasive grain in the surface of the material, and a chip to be formed. The plowing energy is that which upsets the surface material on either side of the abrasive grain as it plows down the arc of cut. The plowing action of an abrasive grain is much like a V-hull boat as it moves through the water. The boat does not remove the water but merely displaces it on either side of its V-shaped hull. The rubbing energy is that which is caused by the rubbing of flats on the surface of the grinding wheel periphery. The wear flats may be due to worn grain (this is termed an attritious wear flat), worn-down bond material, or workpiece material which has stuck to the periphery of the wheel (this is called wheel-loading). It is the rubbing energy which is the most detrimental. Rubbing energy is the energy generated from frictional heat. Virtually all of the heat generated in rubbing is conducted into the surface of the workpiece, increasing its temperature and maybe even changing the metallurgical structure. Even the slightest localized increase in temperature at the surface of the workpiece might cause very sensitive materials to crack. High temperature gradients in the surface of ceramic workpieces are the chief cause of workpiece chipping. Ceramics also suffer from cracking. In extreme cases,

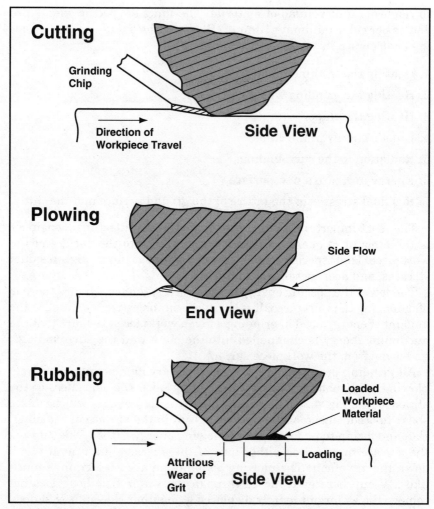

Figure 4.2 Three modes of grinding energy.

where the grinding wheel has become dull and the temperature gradient at the surface is very high, ceramics, and even very hard tool steels like CPM T15, might explode due to localized thermal stress.

Workpiece damage may be minimized by controlling the amount of rubbing energy. This can be accomplished by controlling the wheel sharpness. Continuous dressing controls the rubbing energy, better than any other technique, by keeping the partition of the cutting, plowing, and rubbing energies constant throughout the life of the grinding wheel. Proper dressing and preparation of the wheel periphery is critical, particularly when grinding sensitive materials. With continuous dressing being the exception, no matter what abrasive and bond system are used, the grinding wheel periphery will gradually degrade throughout the grinding process, continually changing the balance of the cutting, plowing, and rubbing energies. It is therefore important to recognize that *all* abrasives, if they are not being continually dressed, will degrade at some rate, requiring the constant performance monitoring either by a skilled operator or an electronic sensing device linked to a decision-making processor, which can activate a corrective function. It is also well to note that the wear-flat area of the grinding wheel governs the produced surface finish. A flat dull grain will remove some material, but, due to its very large, flat area, will rub and burnish the workpiece surface, producing a smoother finish than that of a very sharp aggressive grain, which efficiently removes material, but leaves a surface much like a plowed field with well-defined peaks and valleys. The condition of the grinding wheel periphery and the process speeds and feeds control not only the thermal energy, but also the generation of a surface finish.

Depending on the stock removal rate, the grinding process will use a certain amount of energy, based on the amount of stock being removed, which will have to be dissipated in the arc of cut. It is important to remember that the area of contact between the grinding wheel and workpiece is the only area available over which heat from the grinding process may be dissipated. It is a general misconception that a high horsepower grinding machine can machine faster than a low horsepower machine. This is not necessarily so. If a grinding machine, of say 15 hp, is machining a workpiece to the maximum amount of energy which can be dissipated over the area of the arc of cut prior to a breakdown in the energy balance, then a 30 hp machine cannot machine the part twice as fast. If two workpieces can be fixtured side-by-side, however, then the 30 hp machine can grind two pieces in the same time it took the 15 hp machine to grind only one. In grinding two pieces side-by-side, the effective contact area between the grinding wheel and the workpiece has been doubled, allowing twice as much energy to be dissipated. This also results in an appreciation of the power requirements in grinding. If the width of the grinding wheel in contact with the workpiece is increased, then the amount of power required to make that grind will increase in direct proportion, i.e., double the grinding wheel width of cut and expect the power requirement to double.

There have been many studies carried out on the thermodynamic relationships in the arc of cut of the grinding process. Without fail, each study has related back to the 1920s work on a thermodynamic model of a moving heat source by Jaeger. This work is a classic theoretical study, which mathematically models the heat transfer into the surface of a solid from a heat source moving over it. The practical upshot of a great deal of mathematics is that if a heat source, e.g., a welding torch, were held over a wooden tabletop and moved very slowly across the surface of that tabletop, then a very badly burned and charred surface would result, proving that a great deal of heat was transferred into the surface. However, with the same torch, if it were moved very quickly across the tabletop, the result would be that it would barely leave a scorch mark, proving that only a small amount of heat was transferred into the surface. If the torch were moved very quickly back and forth across the table, along the same line, the heat would gradually build up in the surface of the table until it, too, became badly burned. There is indeed a relationship between the intensity of a heat source and the speed with which it is moved over a surface and the number of times it is moved over that surface. This is very important in our understanding of the grinding process. It can be seen from Jaeger's model that by increasing the speed of the workpiece, the amount of heat transferred into the surface will decrease. From the micro-milling analogy, it was shown that by increasing the workpiece infeed rate, the grinding wheel will act softer, therefore self-sharpening might occur and the process may well be improved. The great majority of grinding processes require the grinding wheel to make a number of passes over the surface of the workpiece before the final dimension is achieved. It would seem safe to assume that heat will build up in the surface of the material and that the greatest amount of heat is generated where the grinding wheel contacts the workpiece surface. Depending on the condition of the grinding wheel periphery, the heat transferred to the surface will change. If a grinding wheel periphery could be prepared to be absolutely sharp, with zero wear to the flat area, then about 95 percent of the energy (heat) generated in shearing will go into the chip and 5 percent into the surface of the workpiece. Of the energy used to plow through the surface of the material, virtually all is conducted into the workpiece. As the grinding wheel wears and grows wear-flats, rubbing energy occurs. Of the rubbing energy, about 95 percent is conducted into the surface of the workpiece. The rubbing energy is the most detrimental to the workpiece surface as it generates so much frictional heat. However, a compromise has to be made with respect to the surface finish required, as it is the wear flats that smear and smooth out the surface material. This additional information has further developed our understanding of another part of the grinding process. If a great deal of stock is to be removed from a workpiece, then a very sharp wheel periphery will ensure that the material will not be damaged, as almost all the heat will be carried away in the swarf. If a high surface finish is required, a dull wheel could be used. However, due to its propensity to generate enormous amounts of heat in the surface of the material, only small depths of cut can be taken.

The question of heat and energy generated in the grinding zone led a number of researchers to explore the bounds of the process and investigate the outer limits of exactly how much heat can be dissipated in the arc of cut. As we have seen, it will depend on how that heat is generated. Heat that is transferred to the workpiece surface increases the temperature of the cutting zone, and therefore the temperature of the cutting fluid.

A limitation on heat transfer occurs in water-based fluids, for example, when the water boils. When the water is a liquid it can transfer heat by convection and conduction, but if the water is heated to a point where it becomes steam, a gas, then it can only transfer heat by radiation. Another aspect of our process model is the limit to the amount of heat which can be generated in the arc of cut; that is, the amount of heat which is transferred to the cutting fluid and changes its state from a fluid to a gas.

Measurements of the causes and effects in the grinding process could themselves be processed using a computer algorithm to close the loop and provide us with a controlled and predictable abrasive machining process. The causes are the myriad of changes we can make in machine tool selection, feeds and speeds, grinding wheel selection, wheel preparation, and cutting geometry. The most sensitive in-process measurement is the normal force between the grinding wheel and the workpiece (see Fig. 4.3). This is the most direct indication of how easy or difficult it is for the abrasive grain to penetrate the surface of the material. If the wheel is sharp, then the grain will penetrate the surface of the material easily and the normal force will be low. Grinding will therefore be taking place in a cutting regime and the majority of the heat will be partitioned off into the swarf. If the normal force is high, then wear-flats have developed, making it more difficult for the abrasive grain to penetrate the workpiece surface and causing the wheel to generate more frictional heat. The normal force can be used as an in-process feed-back mechanism to control the grinding wheel sharpness and automatically initiate wheel dressing cycles on CNC machines.

There are many devices which can measure the power or current draw from the spindle motor, but the grinding power is affected mostly by the tangential or cutting force. Electrical measurements of the spindle power include variables such as motor efficiency, windage losses, bearing friction, etc. As the tangential force is the least sensitive to changes in grinding process, and the least responsive to energy partition changes, electrical current and power monitoring are not good measurements to make. Power monitors will provide a reaction signal to changes in the grinding process. However, there are thermal changes which will affect the surface integrity of the workpiece, which

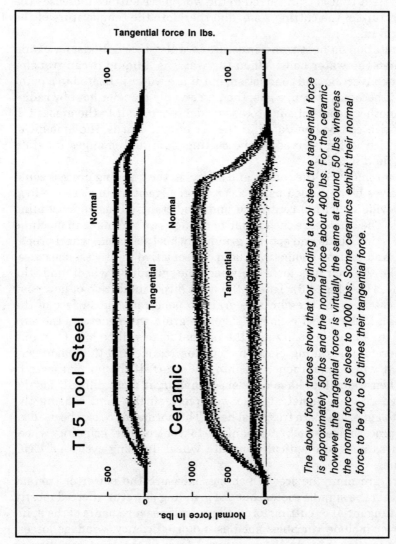

Figure 4.3 Comparison of grinding forces—ceramic vs. metal.

can change quite markedly and very rapidly and will not show on a power monitor, so be careful of such devices, understand what they are measuring, and how you propose to use the data.

There have been countless attempts to close the loop on grinding, none of which have truly reached the goal. Grinding process changes, which have been measured to occur within the space of microseconds, will drastically affect the integrity of the process as well as impart a condition which will set the scene for the next phase of grinding. If it were possible to measure these rapid changes buried in a sea of vibration noise and extraneous signals, then a control system would have to be able to process the signal, decide what change to make, and then make that change in microseconds, while all the time still monitoring the process. Even with the sophistication of super-fast computers and Programmable Logic Controllers (PLCs), there is a sluggish mechanical link between the measurement reaction and the machine tool action. Beware of Adaptive Control (AC) claims. There are force measurement adaptive controls which are purely for dimensional control, particularly in ID grinding and some process AC. At best, they run to a set level of power or force, with a large tolerance on the trigger levels, so that the expected optimum grinding conditions are working to a narrow algorithm, preset to ensure that a certain event, e.g., spindle motor stalling, will not occur. This is *not* adaptive control. Be careful!

Chapter 5

Grinding Machine Tool Design

5.1 Introduction

Grinding machine tools have not evolved, over time, in a manner which allows the new technologies in abrasives and materials to be properly and economically utilized. New and advanced technologies like creep-feed grinding, along with high-speed grinding and the use of super-abrasives, demands higher stiffness and better thermal and mechanical stability, as well as more precise control of the machine tool movement.

Abrasive machining processes are becoming more prevalent in the manufacturing industry and will continue to do so into the next century. Since the early 1960s, abrasive machining processes have been developed to such an extent that they now pose a threat to the traditional metalworking processes of broaching, milling, planing, hobbing, gear cutting, and even turning. A great deal of the manufacturing industry, in its skepticism, is slow to embrace this situation and continues with the outdated, tried and failing, traditional approach to metalworking. A new era awaits in the form of the latest space-age alloys, cermets, plastic composites, and ceramics. These materials are proving to be not only economically difficult to machine, but also technically difficult to machine with respect to surface integrity and workpiece quality. Grinding these materials has proven to be the only way to successfully machine them technically as well as economically. Ironically, the great majority of the world's grinding machine tool builders have neglected to see this potential and their machine tool developments are traditionally lagging behind the innovative materials science. The proper design of a grinding machine is what allows the industry to take full advantage of the latest technologies. Most of the world's machine tool builders fall a long way short, forcing the users to compromise their approaches to modern abrasive machining.

5.2 An Historical Perspective

Since its inception, precision grinding has been regarded as a process reserved only for the light finishing and the precise grinding of materials which exhibit high surface hardness, yet require superficial stock removal. Thus, the grinding machine was designed with that philosophy in mind.

The first surface grinding machine tools were of a cantilever spindle design, intended to provide easy access to the work area and allow convenient changing of the grinding wheel (see Fig. 5.1). The spindle bearings were light-duty, as the grinding process was not a heavy-stock, removal rate process. The machining forces on the system were very low. Grinding was seen to be the very precise machining of material by the removal of extremely small amounts of stock. The sole purpose of the spindle bearing was to run accurately and precisely at relatively high speeds, approximately 30 ms^{-1} (6000 sfm) wheel speed, which for a 150 mm to 200 mm (6 to 8 in) diameter grinding wheel is approximately 3250 RPM, which are certainly higher speeds than those for a typical milling or turning machine of the day. Hence, the bearing preloads were light and, in general, the machines were quite flimsy. As the years progressed and materials science advanced, steel technology progressed, giving us higher strength alloy steels and tougher and harder materials. These newer materials were more difficult to grind because as the grinding forces increased, part size also increased, requiring both larger and stiffer structures. So the grinding machine, still in its basic form, simply grew larger, heavier, and somewhat stronger. Strength typically came from the addition of more material to the structure. As the grinding machines became larger and wheel diameters increased to 400 mm (16 in) and even 600 mm (24 in), the wheel peripheral speed remained the same. Spindle speeds were now decreasing, in the range of 1,450 to 950 rpm, respectively. This decrease in RPM and increase in mass poses a challenge to the machine tool designer to design a structure which is vibrationally stable.

Vibration causes major problems in the traditional grinding process as the grinding wheel depth of cut is so small that even the slightest amplitude of vibration can have dramatically damaging effects on surface finish, wheel wear, and form holding. For a grinding wheel depth of cut of 0.025 mm (0.001 in), the system might see a resonant vibration amplitude of 0.0025 mm (0.0001 in). This is a 10 percent change in wheel depth of cut every wheel revolution. It is usual to find that the

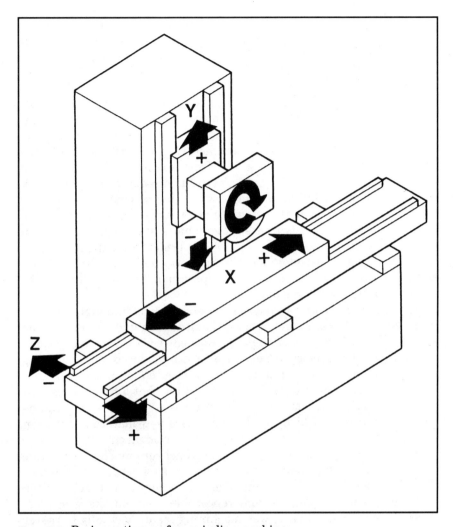

Figure 5.1 Reciprocating surface grinding machine.

major cause of vibration in grinding is due to grinding wheel out-of-balance. For an early machine running at 30 ms^{-1} (6000 sfm), the out-of-balance frequency was 54 Hz (3250 RPM divided by 60 to give cycles per second = 54 Hz). As wheel diameters increased to 400 and 600 mm (16 and 24 in), the out-of-balance frequency decreased in the range 16 to 24 Hz (950 and 1450 RPM respectively) for the newer and larger machines. The machine tool builders faced a dilemma. The addition of material and therefore mass to the structure, for extra stiffness, lowered the natural frequency of vibration for the structure. As we have witnessed, the wheel peripheral speed remained the same, so the forcing, out-of-balance frequency also decreased. Hence, the machine tool builders had to compromise their designs, balancing the increase in stiffness by the addition of material, with structural damping features in the hope that the regions of vibrational instability could be minimized.

For fifty years, grinding machine tool design, through the 1970s, was, at best, a rehash of the apparently "proven" concepts of a manufacturing industry from a bygone era. It is largely due to poor machine tool designs that grinding, as a manufacturing process, has a bad name in the industry. The industry suffers from excessively long machining cycle times on a part with significant value already added. The workpiece is prone to grinding damage by burning or chatter marks, and production supervision is antagonized by the operator's "trial and error" approach to feeds and speeds. This state of affairs began to change when the materials science of the 1960s forced the industry to find better machining methods. The industry had to be far more conscious of not only the dimensional precision but also the surface integrity of the workpiece material.

During the 1960s, the aerospace industry took off into high gear with materials which were much more difficult to machine by conventional means. The superalloys of the day were so resistant to conventional machining with cutters that even the best high-speed steel and tungsten carbide cutters suffered under very high tool wear and very slow feeds and speeds. It was at this point in time that a new process was born out of grinding technology. The process was creep-feed grinding. Here was a process which could attack these new materials with very little problem compared with using the conventional cutters of the day. It was found that by using the grinding wheel very much like a milling cutter, taking a large depth of cut at a very slow feed, even the most difficult superalloys were machined more economically than by cutters and allowed a very precise and repeatable form on the workpiece. Indeed, creep-feed grinding was machining faster, virtually burr-free, and more economically than milling, yet with the precision expected from a grinding process. The success and acceptance of creep-feed grinding since the early 1960s has caused it to become one of the fastest growing processes in the metalworking industry, but this has only come about through grinding machine tools specifically designed with the feeds and speeds and the level of grinding forces which occur in creep-feed grinding in mind.

5.3 Creep-feed Grinding Machine Design

Unfortunately, creep-feed grinding, which developed from the surface grinder principle, has held on to the surface grinder image. It is quite difficult to tell a creep-feed grinding machine from a standard surface grinder. However, for creep-feed grinding to work effectively, a more rigid grinding machine needed to be built. It had to be a grinding machine which had the rigidity of a milling machine, yet the precision and control of a grinder.

The stiffness or rigidity of machine tools is measured in terms of a deflection for a given force applied between the grinding spindle nose and the work table. Typically the machines of yesteryear ranged from 4 to 15 MNm^{-1} (25,000 to 85,000 lbf/in), and the latest creep-feed grinding machines lie in the range 20 to 80 MNm^{-1} (115,000 to 450,000 lbf/in), which is five times more rigid than the conventional grinding machines, although the machine configuration is the same; a cantilever spindle with rectilinear slideways provides an infeed motion, a cross-feed motion, and a traverse motion.

The 1980s brought about the next revolution in grinding systems. Sparked by the advent of continuous-dress creep-feed grinding, a series of automated grinding cells for the aerospace industry appeared. The industry witnessed a turnaround in its perception of grinding as a bad process. No longer was the process itself a problem; grinding cycle times were so fast that the efficient and timely manner in which the workpiece was loaded and unloaded from the machine became the overriding factor. Automation then surrounded these systems, and the marvel of it all allowed the machine tool builders to neglect their fundamental duty once more. Amazingly, the basic machine tool design of the latest multimillion dollar installations from Europe still shows the traditional cantilever spindle and rectilinear slideway arrangement (see Fig. 5.2). Even worse, the addition of an over-the-wheel dressing system to the standard machine configuration, necessary for continuous-dressing, increases the mass of the wheelhead significantly. The latest grinding machine concept is to hang a massive wheelhead and rotating grinding spindle (vibrating in the range 15 to 45 Hz), along with an over-the-wheel dresser assembly (vibrating in the range 45 to 95 Hz), on a cast iron or steel weldment column, with a vertical ball-screw drive supporting the whole assembly, having a travel of 510 mm (20 in) or longer. The precise control of the movement of this system, and the mechanical stability over the range of vibrational frequencies, is a designer's nightmare. The machine tool industry showed little regard for the basic principles developed from the research which brought about the concept of continuous-dress creep-feed grinding. In fact, the research test rig bears no resemblance to a traditional surface grinder, and its features were purposefully designed to overcome the deficiencies of the traditional machine tool. So-called conservatism, which attempts to hang on to the comfortable and traditional approaches of the past, is the major obstacle to progress in the metalworking industry.

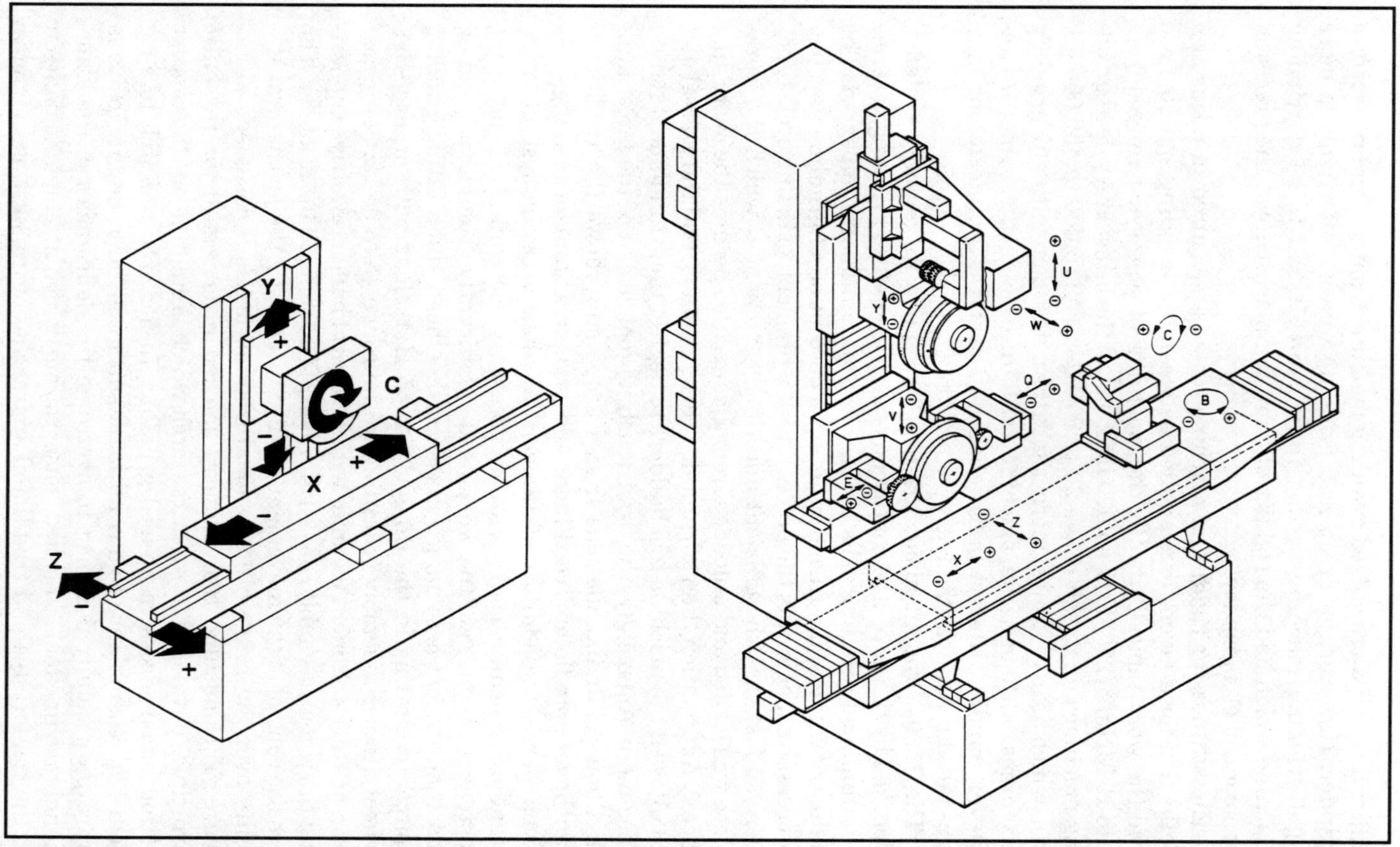

Figure 5.2 Single wheel and dual spindle surface grinding machines.

As we trace the evolution of the grinding machine, it becomes apparent that the machine tool builder is satisfied with the status quo, or, perhaps more to the point, is centered around designing a recognizable machine, which is based on the historical perception of what a grinding machine looks like, rather than how one must be mechanically.

5.4 High-speed Grinding Machine Tool Design

Superabrasives and high-speed grinding are continuing to revolutionize the abrasive machining industry. Since 1969, when CBN first became commercially available, the grinding machine tool industry has wrestled with the design of machines best-suited for superabrasives. Very quickly, initial testing showed that higher peripheral wheel speeds would be advantageous when using CBN. Early on, the speed range was predicted to be in the region of 60 to 120 ms^{-1} (12,000 to 24,000 sfm). Due to the high cost of the CBN grain, smaller diameter wheels would be used; 200 mm to 405 mm (8 in to 16 in). This puts the out-of-balance frequency at anything from 50 to 190 Hz. Naturally, the machines of yesteryear were not designed for these speeds and vibrated badly. The excessive vibration caused very poor surface finishes, chatter, and heavy wheel usage. Peculiarly, when the industry experienced these failures on its antique machines, it decided that CBN simply did not work! It never questioned the basic machine tool design until Japanese and German grinding machines began to appear on the market with the slogan, "Designed to use Superabrasives." Today there are a number of very specialized production grinding machines running at 175 ms^{-1} (35,000 sfm). In the laboratory, test rigs have approached and gone beyond 340 ms^{-1} (67,000 sfm), the speed of sound. At such high speed, the abrasive machining process enters a realm where the conventional approach to machine tool design and engineering is inadequate. Grinding spindle designs have had to be changed to run at higher speeds and to run cool with high stiffness and good vibrational stability. The industry is beginning to see magnetically levitated high-speed spindles with the ability to run 20,000 rpm at 30 kW (40 hp). Such bearings are manufactured by the company IBAG in Switzerland.

Not only was the basic machine tool design inadequate, but also the dressing systems were unsuitable for both the resin and vitrified superabrasive wheels. At ultrahigh-speeds using plated wheels, even the machine guarding has to be redesigned. The bursting forces, and therefore the force from any rotating part which is let loose, increases with the square of the increase in speed. If the surface speed increases by a factor of 4, from 30 ms^{-1} (6,000 sfm) to 120 ms^{-1} (24,000 sfm), then the force increases by a factor of 16.

5.5 Vibration in Machine Tools

The effects of vibration in machine tools are of paramount importance in order to understand the process, as well as assist in trouble-shooting. A fundamental, physical relationship in mechanical structures is that the natural frequency of vibration is proportional to stiffness and inversely proportional to mass. The natural frequency of vibration is the frequency at which a structure will vibrate given an initial impulse force. This is illustrated perfectly on the strings of a guitar. The highest notes are made by plucking a very thin, light string, which is highly tensioned and very stiff. The lowest notes are made from a thick, heavy string, loosely tensioned and flexible. As a structure is made stiffer, the natural frequency increases, and as the mass of a structure increases, the natural frequency decreases. The natural frequency is the dominant, highest amplitude frequency at which a structure will vibrate. There are other harmonic frequencies of vibration which have a significant amplitude of vibration, but none as large as the natural frequency. A machine tool will generally be designed to be stable during a given band of operational frequencies.

High-speed grinding, using superabrasives, is undoubtedly the next generation grinding process. Machine tool designs have to take into account the vibrational stability required and the stiffness necessary, not only to perform the grinding operation, but also to dress the grinding wheels in the proper manner. The great majority of grinding machines today are sadly inadequate in both of these areas. There are partial solutions in the forms of new materials for machine tools, like "Granitan," a patented, epoxy, granite material, which is used to manufacture machine tool bases, and which exhibits extremely good damping properties, coupled with good thermal and mechanical stability. Granitan is the trade name held by Studer in Switzerland, who owns the exclusive rights to manufacture the material. Increasing the stiffness of a structure by designing complex weldments is more prevalent today, as stiffness is improved and extra damping is provided in the design of the welded joints (see Fig. 5.3).

Thermal stability of a grinding machine is essential for super-precision grinding. Changes in the environmental temperature can drastically change the dimensional stability of the machine tool. The work environment can be air-conditioned and maintained at a constant temperature, however, there are significant heat sources in a machine tool which need to be controlled or at least directed. There is heat generated from hydraulics, fan motors on spindle drives, fans in control cab-

Casting Cross Section

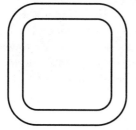

A casting is a homogeneous structure which, like a bell when given an initial impact force, will transmit the resulting vibration throughout the structure and cause the amplitude of the vibration to linger.

Weldment Cross Section

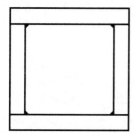

A weldment can be constructed without regard for the foundryman's draw angles, etc. The structure can be designed with maximum rigidity. Unlike the casting, which is homogeneous, the weldment is like a cracked glass when it comes to transmitting vibration. The weld joints act as barriers to the transmission of vibration.

Figure 5.3 Casting and weldment cross sections.

inets, and fans in cutting fluid refrigeration systems, all blowing air around the machine. When running the spindle, bearings increase in temperature and grow to a stable operating temperature. The temperature of the workpieces and fixtures have to be stabilized. Even the operator gives off heat. The human body gives off approximately 100 watts of radiated heat. Imagine the effect on a new super-precision grinding machine of a whole entourage of both interested and uninterested personnel looking on and radiating 100 watts of heat each. Thermal stability has to be designed into the machine, as well as into the placement of the equipment in the workplace.

High-speed grinding and creep-feed grinding have been explored together in an array of materials from metals to ceramics. Again, the trend is higher stiffness, better vibrational stability, and precise control, but another area looms over the traditional machine design. The application of high-speed and creep-feed grinding generates swarf, in a volume far greater then ever before. Conventional machine designs cannot cope with the huge volume, nor can they cope with the high flow rates of cutting fluid, without leaking it into the atmosphere and, more commonly, all over the floor. The velocity of high-speed swarf is easy to overlook, as it was never important in the past. Grinding swarf, travelling not at 60 to 70 mph, but now at 250 to 600 mph, does an extremely fine job of eroding machine guards, workpiece fixtures, and, in fact, anything that is in its path.

For too long we have neglected the basic fundamentals of machine tool technology which will take us into the new manufacturing environment.

5.6 The Next Generation Grinding Machine

There is a need to develop a next generation grinding machine tool based upon the research and development which has taken place over the last 25 years and the direction materials science is taking the manufacturing industry. Some of those features are highlighted below, and may guide us toward such a machine and, at the same time, show us the shortcomings in our existing system.

1. Machining at high-stock removal rates, with high-precision, and in very hard, difficult to machine materials, requires a very stiff machine with accurate and well-maintained bearings.

2. To achieve good surface finish, economical wheel usage, and overall workpiece integrity, a vibration-free system is essential. Also, the system has to be precisely controlled by a motion system, which is both mechanically and electronically stable.

3. High-speed and high-power spindles are required to properly use superabrasives. Proper wheel dressing and wheel preparation systems are essential.

4. A quick and accurate method of wheel changing is required when continuously dressing conventional vitrified wheels, or when a variety of plated-form, superabrasive wheels are being used.

5. As stock removal rates increase, there is a need to cope with large volumes of swarf and cutting fluid. Cutting fluid management is necessary in both its application to the grind and the filtration and conditioning afterwards.

6. As the actual cut time decreases, there is a need to be able to load and unload workpieces out-of-cycle to maximize the machine's efficiency and productivity.

7. Because of the majority of parts machined in the industry today, and probably in the future, it is more stable and easier to manipulate the workpiece and leave the grinding wheel stationary.

Having identified the areas on which to concentrate the effort for the design of the next generation grinding machine (NGGM), it makes a clean paper approach quite simple:

Stiffness, stability, and the ease of changing the grinding wheel quickly have been addressed in a system registered under a 1987 U.S. patent. Using the system of hydrostatic bearings and "squeeze-films," a grinding wheel is held in a doubly-supported yolk, which does not have to be dismantled to change the grinding wheel. The result is a grinding wheel head of exceedingly high stiffness, which sacrifices none of its inherent rigidity to effect a wheel change. This bearing system has a calculated stiffness in excess of 1 GNm^{-1} (6,000,000 lbf/in), an order of magnitude more rigid than the best creep-feed grinding machines today. A similar design of grinding wheel bearing has been built in Japan and has a measured stiffness of 2.6 GNm^{-1}.

Rather than attempting to control the movement of the rotating body of the grinding wheel toward the workpiece, the wheelhead will remain stationary. It will be the function of a part manipulator to position the workpiece with respect to the grinding wheel (see Fig. 5.4).

Bearing in mind the volume and velocity of the swarf and cutting fluid, it is advantageous to grind in a vertical direction, so that the swarf and cutting fluid are blown down and away from the work fixture, keeping the area free from contamination. This is very much unlike the traditional surface grinding machine, where the swarf and grinding grit drop down onto the machine table, heavily contaminating the workpiece, fixture, locators, and clamping mechanisms. This idea was first used on a prototype creep-feed grinding machine and cell system manufactured by the Snow Company in Sheffield, England.

Depending on the type of fixture and control system, the part manipulator can perform surface, plunge form, contour, cylindrical, or cam grinding, all on the same machine tool (see Fig. 5.4). Computer Numerical Control (CNC), has brought a control to grinding which was never there before. The computer control can monitor wheel diameter and compensate the motor speed to maintain a constant peripheral wheel speed. The control can move the axes of the machine to contour a surface with a constant speed of the wheel traverse over the piece part. CNC allows continuous-dressing to be conducted easily and efficiently. Process algorithms can be written in the computer code to insure constant grinding conditions for every part and from part to part.

A major problem today is getting parts in and out of the machine fast enough. A modern-day modification to the design of the yesteryear machine is the combination of a moving column and rotary table. The

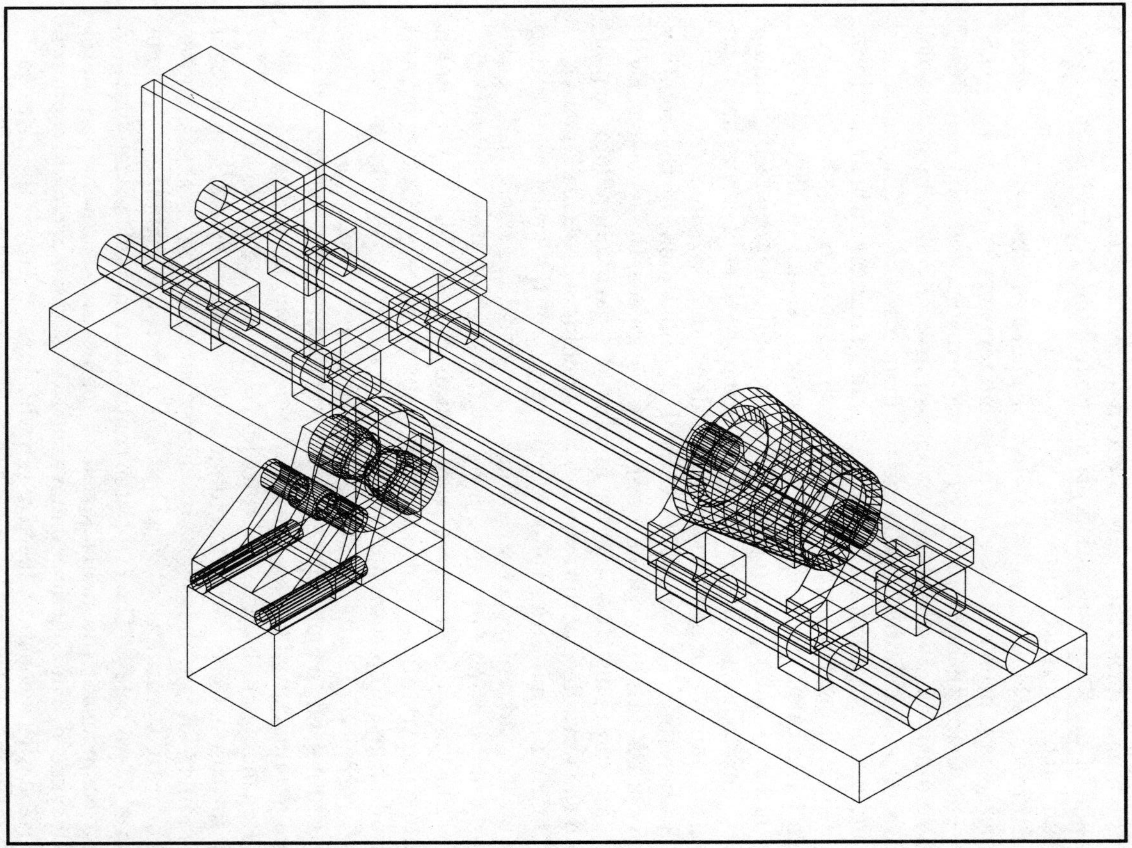

Figure 5.4 Surface and cylindrical NGGM concept.

Grinding Machine Tool Design   115

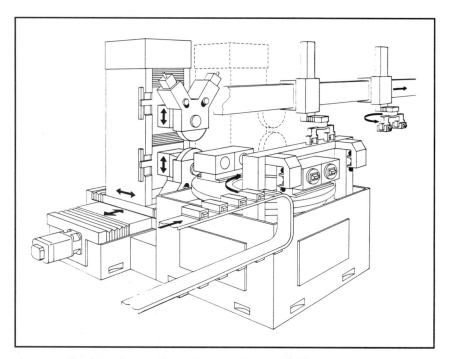

Figure 5.5 Combination moving column and rotary table machine.

premise here is (see Fig. 5.5), with a large enough rotary table, parts can be loaded and unloaded at 180 degrees to one another, affecting an in-cycle load and unload operation. Unfortunately, for this system to work, the wheelhead, along with whatever else has been hung onto the machine column, and its vertical ways have to be transported by a horizontal ball-screw along a horizontal way system. As the column moves along the horizontal ways, the error in those ways rocks the column as it traverses the workpiece. Moving column designs magnify way errors and suffer from the settling error in the rotary table after it indexes its 180 degrees. Notice that the rocking column and the rotary error affect the grinding process in the direction of the wheel depth of cut. By using the rotary axis with a vertical travel machine, the error is in the direction of the traverse and not the depth of cut.

Such a NGGM system is being developed for surface, external diameter, and contour grinding, using both high-speed and creep-feed. The machine may be a completely de-coupled system (see Fig. 5.6), isolating the wheelhead from the part manipulation system, or a coupled unit, as in more conventional machine tools (see Fig. 5.7). The stationary head aspect of the design is bringing the most, in terms of accuracy and stability, to the NGGM system, yet, generally, because of its "non-traditional" look, raises the most eyebrows. However, it is odd that

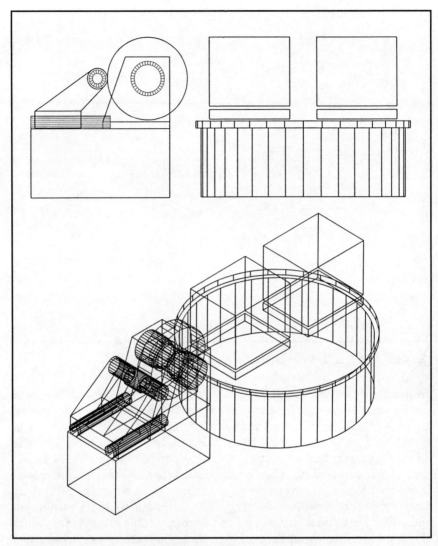

Figure 5.6 De-coupled, rotary table NGGM.

when we look back to the dawn of our industrial age, we were doing just that; the knife grinder used a stationary wheel and manipulated the workpiece across the face of the wheel (see Fig. 5.8). Perhaps we lost something along the way?

5.7 Computer Numerical Control (CNC)

Perhaps the most significant advance in grinding is the increasing use of CNC controls, providing consistency and control.

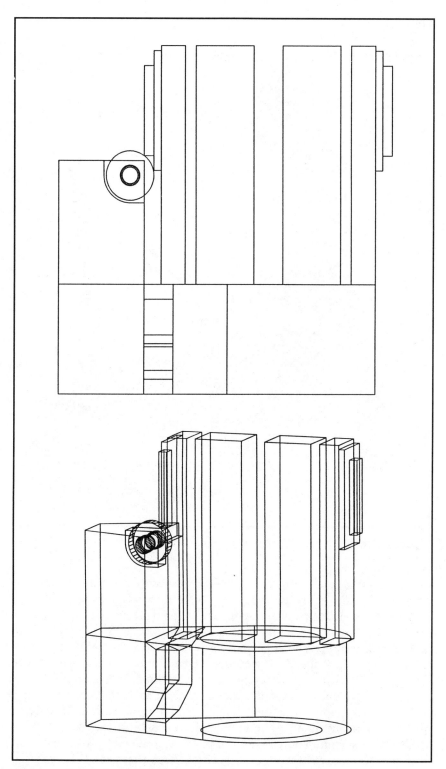

Figure 5.7 Coupled unit NGGM.

Figure 5.8 Knife grinder.

When Numerically Controlled (NC) machines were first developed, the basic means of control was point-to-point positioning in a series of sequenced, rectilinear movements. This was ideal for drilling machines to find the coordinates of a series of holes and then drill them. Milling and turning processes followed naturally, however grinding never enjoyed the NC stage of development. Grinding is mostly a reciprocating process, with process cycles for machining as well as dressing. Grinding does not lend itself to point-to-point logic motion. Even now, with CNC, some machine tool builders find a great deal of difficulty in properly integrating a CNC into a grinding machine design.

CNC is a system whereby a number of machine tool axes can be controlled with respect to velocity, position, and their relationship with one another. The control system can be open-loop or can have a close-loop feedback. Super-sophisticated CNC machines can be controlled via a computer algorithm. In essence, a servo or stepper motor can be controlled by a series of pulses, e.g., a stepper motor might be designed with 1800 steps for every revolution. That means that if the motor is told to step once by the computer, then the motor turns 0.2 degrees (360/1800 degrees). If the motor is linked to a ball-screw with a pitch of 2 mm (0.08 in) per revolution, then for every pulse to the motor, the nut on the screw will advance 2/1800 = 0.001 mm, (0.08/1800 = 0.00004 in). Once the motor moves, a signal can be sent back to the computer by an encoder (a device which counts pulses of angular position) or a linear scale (a very accurate glass scale and light-band sensor which measures increments of linear movement) to verify that the programmed move has taken place. CNC allows more than just rectilinear positional moves to be made. NC was sufficient to make independent rectilinear moves. With CNC a computer can calculate multi-axis positions extremely quickly and so move more than one axis at a time and coordinate each move with respect to the other. For example, instead of moving the grinding wheel across a flat workpiece, we may wish to grind a circular path along the x axis, while rising and falling in the y axis direction. This could be done with NC, by first hand-calculating and then programming every very closely spaced coordinate point in x and y around the arc. The computer can calculate the path for us, making it a simple programming task of telling the computer the radius of the circle, the position of its center, and the start and finish points on the circle in either the clockwise or counter-clockwise direction. The equation for a circle is $R^2 = x^2 + y^2$ where the center of the circle is the coordinates 0,0 in x and y. The computer can maintain the peripheral speed of the grinding wheel by monitoring the position of the dressing axis. If the dresser axis moves 0.01 mm (0.0004 in) then the computer assumes that the wheel has decreased in size by 0.01 mm (0.0004 in) on the radius. It can then calculate the wheel size and update the speed of the spindle motor to increase RPM to maintain wheel speed. CNC, depending on the machine tool manufacturer, can make programming a very easy task, as the operator can be prompted with questions to answer and the computer can work out all of the positional moves. Once the cycle start button is pushed the CNC will perform that task consistently and regularly until interrupted. Couple the CNC with sensory feedback, and the grinding machine can very quickly become self-controlling, requiring only the changing of grinding wheels and the loading and unloading of workpieces.

Some examples of CNC coded lines of information are as follows:

> To start the spindle rotating at 1250 RPM, the line of code might read—M03 S1250. Translated, the code reads, turn on the spindle motor M03

and run at a speed of S1250, 1250 RPM. To stop the spindle, the code might be M05 (M03 meaning turn on and M05 meaning turn off).

There are two types of movement, absolute, G90, or incremental, G91. The machine has to be "homed," that is, all the axes have to be sent to a fixed position where the axes are set at zero. On a surface grinder this may be the table all the way to the left, the cross slide all the way to the back, and the column all the way up. An incremental move is one which is made when the distance is called out: G91 X5. This would mean move the x axis 5 mm (5 in, depending on the metric/imperial unit selector) from where it is. An absolute move G90 X5 would mean to move the x axis to the position 5 mm from the referenced zero position. To complete the line of code a speed has to be programmed, e.g., G90 X25 F250, meaning move the x axis to the position 25 mm from the referenced zero at 250 mm/min. The CNC can make a vectored move, e.g., G90 X50 Y100 F500, which means that the machine will move both the x and y axes from where they are to positions 50 mm and 100 mm from the referenced zero with a speed of 500 mm/min along that vectored line.

Fortunately, the metalworking industry has escaped and managed to get by with the inefficiencies and delinquencies of the traditional grinding machine tool design for many years. The modern abrasive industry, however, is in an enviable position, since the new age of materials, particularly ceramics, whisker-reinforced metals, cermets, and plastic composites, leaves little room for the conventional metal cutting processes. There is fast becoming a situation where CNC grinding is the only way to machine these materials.

Chapter 6

Cutting Fluid Application and Filtration

6.1 The Role of the Cutting Fluid

The cutting fluid performs four main duties: assists in the dissipation of the grinding energy, chemically assists the material removal process, lubricates the cutting action, and washes away the grinding debris from the wheel/workpiece interface. There are three main types of cutting fluid:

1. Gases, most commonly air
2. Water-based fluids, where the great majority of the fluid (over 80 percent) is water
3. Straight oils, where the fluid is 100 percent oil

Water-based fluids are the best conductors of heat, whereas straight oils are the worst. Conversely, water-based fluids do not lubricate the abrasive process very well, whereas straight oils lubricate exceptionally well. Here, therefore, are the reasons why straight oils generally produce a better surface finish than water-based fluids (see Fig. 6.1). A dull wheel, with large wear flats, will rub and burnish the surface of the workpiece. The large flats will be efficiently lubricated by the oil, allowing a high degree of surface finish to be generated by the smoothing action of the flat grains. Oil has a boiling point higher than that of water by approximately 200°C, depending on the type of oil, so it can support higher surface temperatures before thermal breakdown occurs. Water-based fluids lubricate poorly and have a lower boiling point, so they suffer thermal breakdown under lower levels of heat, which is generated in the arc of cut (see Fig. 6.2). This can be illus-

Sharp abrasive

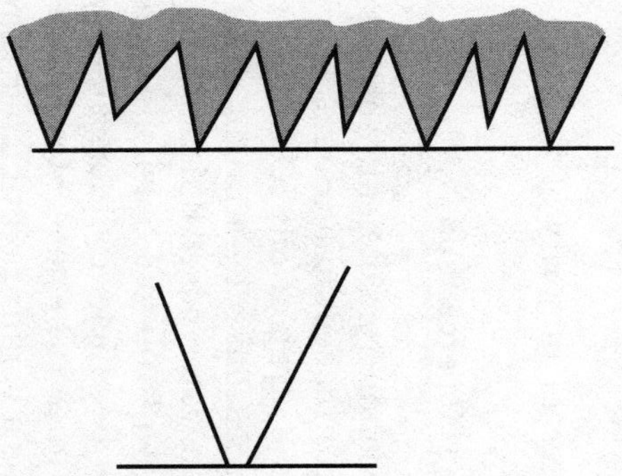

Small "wear flat," easy surface penetration transferring the energy into the chip and away from the surface. Resulting surface is sharp peaks and valleys.

Dull abrasive

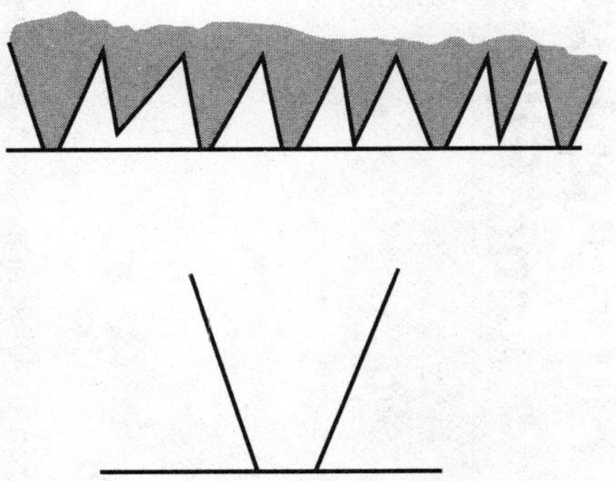

Large "wear flats," lubricated by the oil, generating less frictional heat. Material is removed from the surface with a pronounced smearing and smoothing action.

	Cooling ability	Lubricating ability
Water -	Very good	Very poor
Oil -	Poor	Excellent

* High risk of burning with dull abrasive.
* Allows a dull abrasive to grind with less heat.

Figure 6.1 Sharp abrasive vs. dull abrasive.

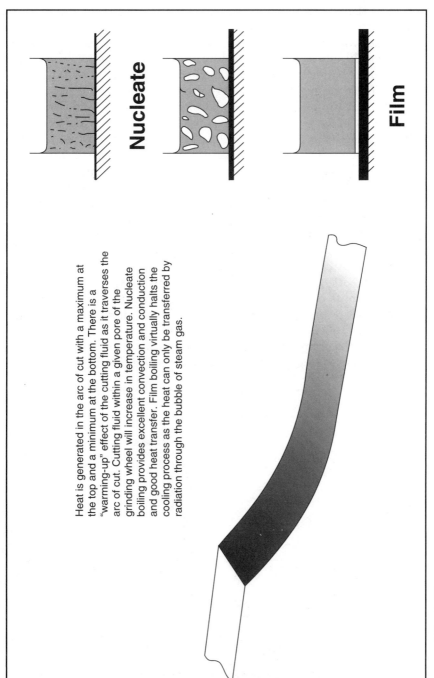

Figure 6.2 Film boiling.

trated by the phenomenon called "film boiling." It can be seen by flicking water onto a very hot stove top. The beads of water will not turn to steam, but will be supported on a film of steam gas, so that none of the heat from the stove top will be conducted or convected into the water. The beads of water will simply run off the stove top, having taken little heat from the surface. In the grinding arc of cut, the heat is generated more at the top than at the bottom of the arc of cut. Therefore, particularly in long arc of cut processes like ID grinding and creep-feed grinding, the cutting fluid warms up as it travels around the arc of cut. It is most important, therefore, to keep the temperature of the cutting fluid constant as it enters the arc of cut. There are a number of heat sources which can increase the temperature of the cutting fluid. There is a tremendous amount of energy released from the pumping action of the system pumps, as well as from the churning of the fluid by the grinding wheel rotation. A significant amount of heat also comes directly from the energy of the grinding operation. A constant cutting fluid temperature is an important control over the process (see Fig. 6.3). If the cutting fluid is allowed to increase in temperature, it will rise closer to its boiling point, leaving little room for the extraction of the heat before the fluid turns to steam, in the form of film boiling, across the arc of cut.

One of the assumptions made, in the application of cutting fluid and the extraction of the grinding energy from the arc of cut, is that the cut-

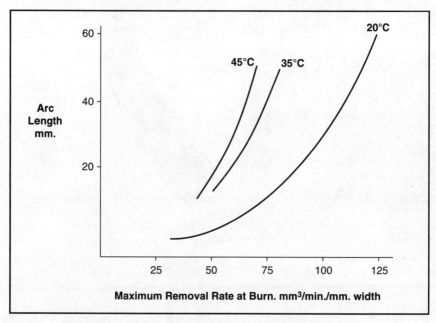

Figure 6.3 The effect of the cutting fluid temperature on the stock removal rate.

ting fluid is indeed applied properly. There is a myth in the industry that there is an impenetrable air barrier around a rotating grinding wheel, which can be broken only by the combination of extremely high-pressure jets and deflector shields, etc. It is simple to dismiss the myth. It is true that there is a turbulent layer of air around a rotating grinding wheel. The surface of the wheel is rough and the surface speed can be in excess of 115 km/h (70 mph). If one takes a table tennis ball and drops it onto a rotating wheel, catching it before it hits another object, there will be scratches on the ball, proving that it must have touched the surface of the wheel. If a very light, spherical object can "burst" through the air barrier, then there is no need for ridiculously high-pressure jets of fluid or vast flow rates. In most grinding applications the cutting fluid is "dribbled" in the general direction of the arc of cut, and, although there are some surface tension effects which can assist in carrying the cutting fluid into the arc of cut, the great majority of the fluid merely cools the surface of the workpiece *after* it has been machined. This is a very bad situation, as cooling the surface directly after it has been machined means that the surface is quenched, causing surface tensile stresses, and initiating cracks and generally poor surface integrity. The key to proper application of the cutting fluid is to apply the fluid at wheel speed. If the fluid is applied to the grinding wheel periphery at wheel speed, then all of the open pores can be amply filled with fluid and transported around the arc of cut. If the fluid is applied in large volumes at anything less than wheel speed, then the wheel will not inherently suck the fluid into its pores. At best, some surface tension effects will drag a very small amount of the fluid applied into the arc of cut, but the rest of the fluid will be blown away by the "spin-dryer" effect of the rotating wheel. Therefore, it is important to manufacture cutting fluid nozzles so that the speed of the fluid is equal to or slightly faster than the speed of the grinding wheel. The reason for a slightly faster speed is that the nozzle will not be positioned close-up against the wheel. Over only a short distance the velocity of the fluid can drop off quite rapidly. A special case arises with very high-speed, plated wheels, where there is no porosity. A manifold and shoe (see Fig. 6.4) is the best method for applying the cutting fluid to the wheel periphery. A large volume of fluid fills the manifold, is dragged up to wheel speed in the shoe, and then is taken into the arc of cut. Nonporous wheels also suffer from hydrodynamic lift or aquaplaning. The oil is compressed in the gap between the grain bond and workpiece surface. As the swarf fills the gap the pressure increases and lifts the wheel away from the surface being machined, creating a dimensional error as well as a change in chip geometry, which will affect the surface finish. Perforations can be made in metal-bonded and plated wheels in the form of slots or radial holes to dissipate the pressure and in some

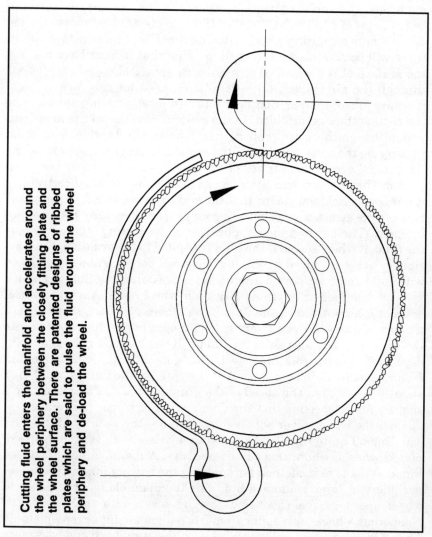

Figure 6.4 Cutting fluid application method for high-speed cylindrical grinding.

instances actually assist in the application of the fluid. The pattern of holes and slots on the periphery of a grinding wheel is critical. Peculiarly, when a series of slots are made in a grinding wheel, they are put in symmetrically and equally spaced, thus forcing a vibration from interruption in the periphery into the system, just like a milling cutter as it cuts into the material with each tooth. Random spacing of slots and holes is preferred in order to prevent a vibration or siren effect. The slots, and particularly the pattern of the holes, will most definitely affect the longevity of the form profile and, of course, decrease the amount of abrasive on the wheel periphery. When problems arise, one of the last things to do is break up the wheel periphery with slots and holes.

Maintaining the flow of fluid through to the end of the arc of cut is also a very important part of fixture design (see Fig. 6.5). As the grinding wheel exits the workpiece, the cutting fluid will splash off the back face of the part and cause starvation of the fluid in the arc of cut. The support dam, built into the fixture, helps to hold a volume of fluid in place through to the end of the cut. It is well to notice that as the grinding wheel exits the cut, there is less bulk of workpiece material which can act as a heat sink. The cutting fluid dam is an essential part of good grinding fixture design. In creep-feed grinding, in particular, where the arc length of cut is long, it is quite evident when a dam is necessary since a ramp will occur at the end of the cut. Heat from the grinding is conducted into the workpiece material as the wheel exits the cut. The heat causes the material to expand into the grinding wheel and the wheel machines the expanded material, which, once it has cooled to ambient temperature, shows up as a ramp off the end of the workpiece. It is important not to confuse such a ramp with deflection in the system. A ramp at one end (the exit end) of the workpiece signifies a thermal problem, whereas a ramp at both ends of the cut indicates a mechanical deflection.

6.2 Filtration of the Cutting Fluid

A question which often arises is how clean must the cutting fluid be for a given application? Generally 8 to 12 micron (0.0003 to 0.0005 in) filtration is adequate for general purpose grinding. If surface finish requirements are critical then filtration to 1 micron (0.00004 in) or less is essential. A rule of thumb is that filtration should be carried out to $\frac{1}{20}$th to $\frac{1}{50}$th the size of the abrasive grain used in the grinding wheel. When using superabrasives in resin bonds, it is suggested to filter to $\frac{1}{100}$th the size of the superabrasive grain used in the grinding wheel. Particulate in the cutting fluid acts as an abrasive on the resin bond of the grinding wheel and, as the amount of particulate builds up in the fluid, a slurrying action occurs which can drastically reduce the life of the grinding

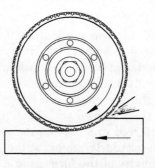

In surface grinding there is an acute angle formed between the grinding wheel and the surface to be machined which assists in the application of the cutting fluid.

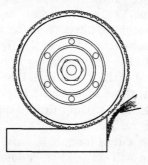

As the back end of the workpiece passes through the grinding wheel, an obtuse angle develops starving the arc of cut from cutting fluid.

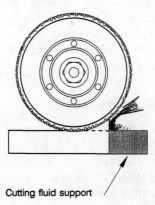

Cutting fluid support

A dam or fluid support, designed into the fixture, can help keep the flow of fluid into the arc of cut through to the end of the cut.

Figure 6.5 Fixture design necessary to maintain good cutting fluid flow.

wheel. It must be realized that filtration down to very fine particulate sizes requires special methods. Magnetic separation is satisfactory for magnetic materials where only a small amount of swarf is to be produced. Paper filtration is the preferred system as it can cope with both large volumes of fluid, as well as very large volumes of swarf. Cyclone and centrifuge type filtration requires a great deal of energy, which generally ends up in the fluid as unwanted heat, and also takes a significant time to achieve good filtration. Diatomaceous earth can be used for extra fine filtration, but requires a great deal of attention to maintain the system. Also, very fine filtration methods can strip the essential additives in the cutting fluid, which are purposely in place to assist in the grinding process. Large volume systems are preferable as they allow some settling of the fluid so entrapped air can escape, reducing foaming and allowing the fluid to cool. A rule of thumb is that for the maximum flow of the system per minute, a tank capacity 10 to 25 times that should be used. For example, if the main cutting fluid pump delivers a maximum flow of 6 liters/s (95 U.S. gpm) then the tank capacity should be between 3600 and 9000 liters (950 and 2375 U.S. gallons).

Refrigeration of the cutting fluid is particularly important in high production grinding, creep-feed grinding, or grinding with superabrasives applications. A refrigeration unit is essential as it can be accurately controlled to within two or three degrees of temperature, whereas a cooler/radiator (see Fig. 6.6) can bring the temperature of the fluid ten to fifteen degrees below the ambient temperature of the air being blown across the cooling fins. Therefore, if the ambient temperature fluctuates then so does the temperature of the fluid. Cooler/radiators also blow enormous amounts of air around the shop environment and generate a great deal of low frequency noise.

Maintaining a constant and cool temperature of the cutting fluid not only is good for the grinding process, but also increases the life of the cutting fluid. Bacterial growth can run rampant through water based cutting fluids. Over a period of only a few days, in a poorly disciplined shop, a cutting fluid can become contaminated with tramp oils, beverages, litter, and waste matter. At temperatures above 28°C (83°F) bacterial growth accelerates and the fluid smells foul, particularly if left to settle over a weekend. The properties of the cutting fluid are seriously depleted, however, the risk of dermatitis and infection in the form of open sores or wounds to the operators is much worse. Poor management of cutting fluids not only is bad for the grinding process, but also presents a very serious health risk to personnel.

6.3 Types of Cutting Fluid

There are four main types of liquid cutting fluids:

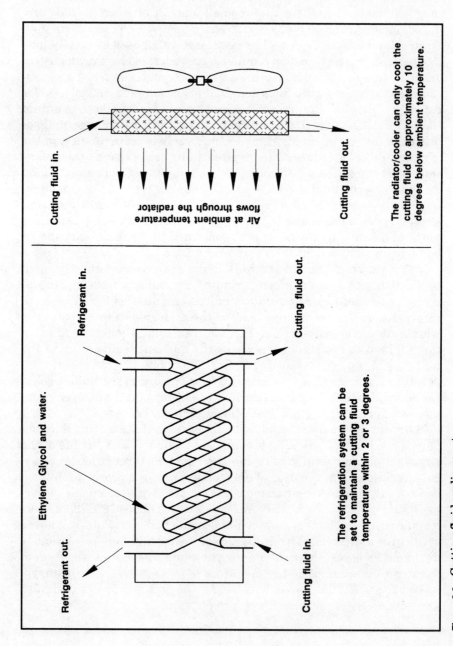

Figure 6.6 Cutting fluid cooling systems.

1. *Straight Oil*—Straight oils, as the name implies, are mostly petroleum-based hydrocarbon fluids, sometimes with additions of animal fat, vegetable oil, or marine oil to improve surface wetting and lubricity. These fluids generally have Extra Pressure (EP) additives to provide even greater lubricity under arduous machining conditions. EP additives are typically sulphur, chlorine, and phosphorus. Straight oils tend to produce the best surface finishes and provide an environment in which superabrasives enjoy an extended life. Care has to be taken in choosing such fluids because some workpiece materials are very sensitive when exposed to some EP additives, particularly at elevated temperatures evident in the grain/workpiece surface interface. Often, sulphur has been the culprit in the hydrogen embrittlement of high-strength alloy steels. Also, there have been medical studies carried out on operators exposed to the atomized particles and fumes from straight oils. The results have shown a high incidence of cancer-related illnesses. Couple the medical findings with oil's propensity to catch fire at relatively low flash points, and it is easy to see why there is great pressure on the cutting fluid research establishments to develop nonpetroleum-based cutting fluids. The fact that oil is, by far, the best lubricator, yet not necessarily a good conductor of heat, makes the task of developing new fluids a major challenge.

2. *Oil/Water Emulsions*—Though somewhat of a misnomer, as oil cannot be completely emulsified in water, these fluids are more than 80 percent water and the rest oil with chemical additions, which assist in the dispersion of the oil droplets in the water. These additions are soap/detergent type fluids. Oil/water emulsions are generally milky in appearance. The ratio of oil to water affects the lubricity of the fluid. These water-based fluids are more able to conduct the heat from the grinding operation than straight oils. However, rich fluids can be quite lubricative, particularly if they contain extra EP additives. A problem generally arises when the oil/water mix becomes too rich and begins to foam under the churning action of pumping and grinding with the fluid. Defoaming agents can be added to the fluid, however these agents will tend to degrade the performance of the fluid, particularly if they are added in large quantities, which will decrease the wetability of the fluid. If a defoaming agent is used, only a minimal amount is required as an interim measure before a more suitable cutting fluid can be found. Deionized water will assist in extending the life of water-based fluids and will reduce the tendency of a rich mix to foam.

3. *Synthetic Fluids*—These too are water-based fluids, however, with no additions of oil. They are purely water solutions of liquid chemical additives. That means that the additions to the water have com-

pletely dissolved. The solutions are usually clear with a hint of color. They do not have as good rust inhibition as fluids containing oil. Most of these fluids are inactive and do nothing more than cool the cutting zone; some, however, are active and contain chemicals which assist in the machining to form surface chemistries, which require much lower shearing force to grind. Such fluids can have an adverse effect on the machine tool itself in the forms of rusting, corroding, paint stripping, as well as attacking certain types of seals, and having no compatibility with lubricating oils whatsoever.

4. *Semi-synthetic Fluids*—These fluids are a hybrid of chemical solutions and the oil emulsions. Therefore, they combine the properties of each. In fact, these fluids have been developed to a point where a 10 to 15 percent solution can have almost the same lubricity of a straight oil, yet have the cooling capability of water. These are generally the premium fluids, and the most expensive, however, for grinding very sensitive materials and for producing the highest quality parts in a relatively safe environment, these are the cutting fluids of the future.

Disposal of these fluids is of major concern as careless disposal can have dire consequences, not only legally for the company, but also environmentally for the future and in terms of health hazards, which affect us all. Some cutting fluids are composed of active chemicals, which can be hazardous to humans as well as wildlife. When disposing of a spent fluid, remember that it is contaminated with material from the grinding wheel and the material which was ground. Workpiece heavy metals, polymers, and plastics will eventually leach out of the fluids, and there may even be a chance of a reaction with other fluids in a dumping area, causing unexpected disaster. There are specific manufacturer's instructions and local byelaws, governing the disposal of cutting fluids, which must be followed. It is a very good idea to follow them.

Chapter

7

Cylindrical Grinding Processes

7.1 Outside Diameter (OD) Grinding

The development of hardened steel in the latter part of the nineteenth century prompted the need for machines capable of finishing workpieces made of materials which were as hard as, if not harder than, the turning tools of that time. One of the first machine tool developments was the cylindrical grinding machine. In its most basic configuration, the cylindrical grinder is defined as a machine for machining round components. A cylindrical grinder consists of the following main assemblies: the machine base, wheelhead, table, headstock or workhead, and tailstock or footstock (see Fig. 7.1).

The workpiece is typically held, and driven between centers, or supported in a center, and driven by a revolving chuck or collet arrangement. Workpieces may be clamped to a face plate to be cylindrically ground, with extra support perhaps, from a center in the tailstock. As the cylindrical grinder evolved it became the means to manufacture precision shafts, tapers, forms, and cams. Precision threads and leadscrews were also manufactured on cylindrical grinding machines, which were equipped with the necessary mechanical gearing to allow the grinding of a helix. Both the grinding wheel and the workpiece always rotate, generally in opposite directions at the arc of cut (see Fig. 7.2). CNC cylindrical grinding machines that produce such precise shapes today are extremely accurate, not only in their manufacture, but also in their control and feedback systems. The cylindrical grinding machine took the manufacturing technology of the day from a crude art onto a new plateau of higher accuracy and consistency of parts produced. Subsequently, the performance of the parts they manufactured improved. The engineering requirement for extreme accuracy and consistency of rounded components followed, particularly in the automo-

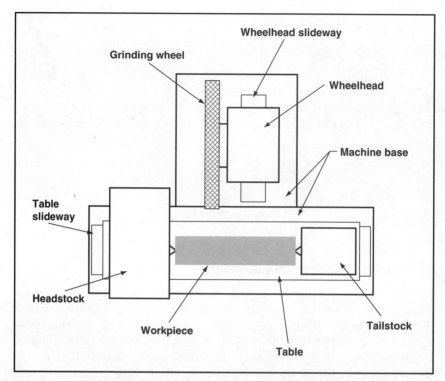

Figure 7.1 Basic cylindrical grinding machine configuration.

tive and aerospace industries. The process quickly became automated due to the high batch quantity of parts to be machined. Especially in the automotive and aerospace industries, many of the cylindrical grinding operations run on grinding machines, which are essentially unmanned for long periods of time, maintaining both the size tolerances and surface integrity of the parts.

Cylindrical grinders are manufactured in many sizes to accommodate a range of diameters and lengths of workpiece. They range from very small machines for job-shop applications, which may be supplied with simple manually operated controls, through medium-sized machines, to high-production machines, which are more automated with CNC and a high degree of sophistication, to extremely large machines, designed for the manufacture and refurbishing of steel mill rolls.

Cylindrical grinders are divided into four main types: plain cylindrical or roll grinders, universal ID/OD grinders, and cam and centerless grinders. The cam grinding process is not necessarily cylindrical grinding, as the cam is not, of course, round. However, most of the principles of cylindrical grinding apply to cam grinding. Centerless grinding will

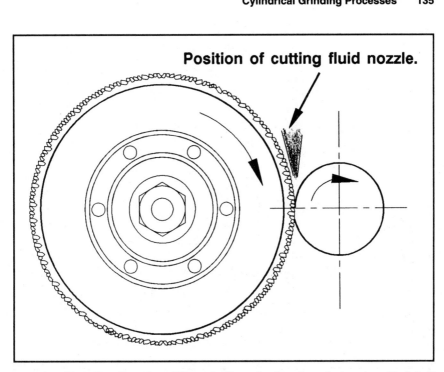

Figure 7.2 Rotation direction of the grinding wheel and workpiece in cylindrical grinding.

be dealt with as a special grinding process, because although the workpieces are ground round, they are not held or fixtured.

7.2 Plain Cylindrical Grinding

A plain cylindrical grinding machine is the most basic configuration. The machine bed is a large and massive structure, giving structural strength as well as both static and dynamic stability to the machining system. As with surface grinding machines, cylindrical machines have been designed using epoxy concrete bases for superior vibrational and thermal stability over the more traditional cast iron bases. The machine base, be it cast iron or epoxy granite, has a guideway, very accurately machined, parallel to the axial motion of the workpiece. The machine table, which is equipped with the headstock and the tailstock, runs on the guideway. The table is split into an upper and lower part. The lower part contains the guideway and the upper part carries the headstock and tailstock. The upper and lower parts of the table may be swivelled about their midpoint to create tapers (see Fig. 7.3). The upper part of this table may be swivelled through 15, even up to 20 degrees, inclusive angle with respect to the main table axis, depending on the machine

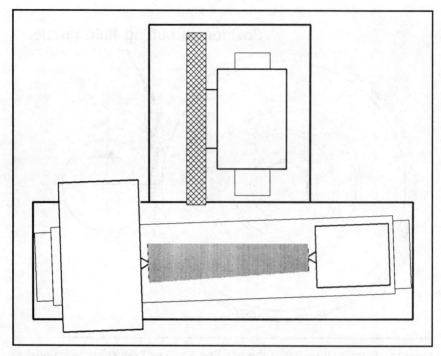

Figure 7.3 The swivelling table allows cylindrical grinding of tapers.

model. This swivelling action allows for the correction of taper in a workpiece and also creates tapers up to the maximum available angle.

The headstock, mounted on the upper half of the table, houses the drive motor and the clamping system for the workpiece. Cylindrical grinding machines have to be rigid and dynamically stable. The concentricity of the ground workpiece depends largely on the quality of the headstock bearings, which are more commonly plain bearings, and the method by which the workpiece is held. It is important to remember to run the headstock for a short time if the machine is cold in order to warm up the plain bearing prior to grinding. The workpiece clamping system might be a face plate, a chuck, a collet, or a dead center with a dog drive (sometimes called a pin and carrier drive). The dog drive (see Fig. 7.4) is always set up so that the pin engages squarely with the dog (carrier). The pin is held in a bracket, which has a left-hand threaded screw holding the bracket onto the face plate. It is very important to ensure that the pin engages with the dog (carrier) so that the pin will pull the dog (carrier) and not push it. The tailstock is the support for the opposite end of the workpiece and might be a live or dead center. The headstock and tailstock may be positioned anywhere along the length of the upper table in order to clamp the workpiece between the centers. The workpiece is always driven.

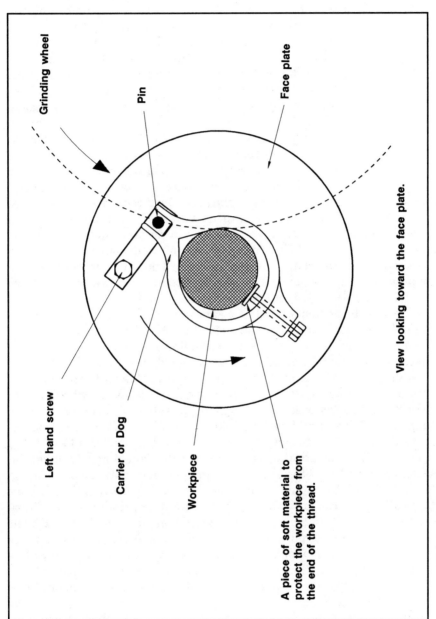

Figure 7.4 The dog drive.

In the case of center held workpieces, there are two types of support: live centers and dead centers. Live centers are centers which rotate with the workpiece. No movement occurs between the center and the center hole in the workpiece. The live center is held in the tailstock, spring-loaded into the center hole in the workpiece, and set to a nominal pressure. Little heat is generated with a live center and no wear takes place between the center and the center hole in the workpiece. This method provides good concentricity and minimal run out. These types of centers are not prescribed for extremely accurate or heavy work, as the run out of the workpiece will bear a direct relationship to the accuracy and clearances in the bearing of the live center. Typically, live centers will produce concentricity within 0.0005 in.

Where very high concentricity is required, and particularly where a part is loaded and unloaded more than once through a series of grinding operations, dead centers are the best. Dead centers are made solid from very hard steel or steel taper shanks with a tungsten carbide cone insert. The center taper hole should be accurately drilled *and* honed to a fine finish for both low friction and high precision. The lubrication of dead centers is essential. Lard oil, soap, or heavy grease is best used for this purpose. Typically, dead centers will produce concentricity within 0.0001 in.

The wheelhead is mounted on a slideway at right angles to the table guideways and houses the main grinding wheel spindle and spindle bearings. A universal grinder has a similar arrangement. In addition to the swivelling of the table, the wheel head may be swivelled to any angular position up to 90 degrees to the upper table for tapers and face grinding. Another hybrid of the cylindrical grinder is one which has its wheel head slideway set at an angle to the table ways so that angular approach plunge grinding may take place. With CNC it is not necessary to angle the wheelhead ways, but merely swivel the head to an angle and interpolate the x and y axes so an angle approach can be achieved (see Fig. 7.5). A universal grinder generally includes an ID spindle. The ID spindle is a most precise high-speed spindle, running typically in the range of 3000 to 175,000 RPM for grinding inside diameter bores. At these very high speeds, ID spindles tend to be driven either off a series of pulleys from the OD spindle motor drive or independently by an air motor or hydraulic motor. A high-frequency electric drive motor is the preferred type of drive motor. The same table swivelling action described above also allows for internal tapers to be ground (see Fig. 7.6).

OD cylindrical grinding is a machining process where a grinding wheel, generally larger in diameter than the workpiece, is used to traverse the axis of the cylinder being ground to form a cylindrical shape. This shape might be a perfect cylinder, a taper cone, a series of diame-

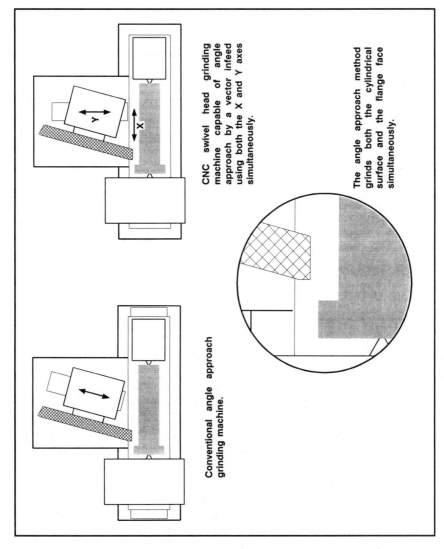

Figure 7.5 Angle approach grinding.

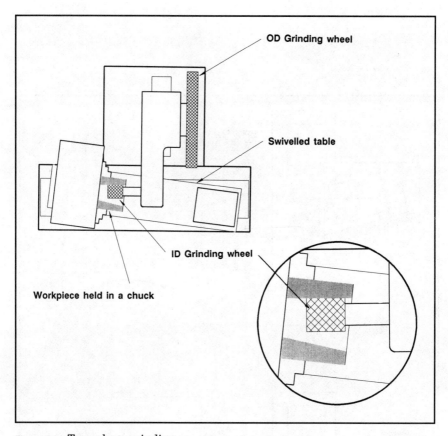

Figure 7.6 Taper bore grinding.

ters, or a combination of all of the above, as well as families of helical forms. In the case of large steel mill rolls for rolling sheet metal, there are cases when the grinding wheel may be smaller in diameter than the roll. Mill rolls, 450 to 1500 mm (18 to 60 in), in diameter, may be plain diameter rolls or slightly crowned at 0.02 to 0.08 mm (0.001 to 0.003 in) across a 2 m (6 ft) length or longer.

Plain cylindrical grinding requires the machine to be set up very accurately. After preparing the grinding wheel periphery, the part is usually ground and then checked for taper. Any taper error can be corrected by a fine scale taper adjustment on the swivel table. On large roll grinders the roll support shoes are adjusted.

Cylindrical grinding feeds and speeds are very important not only for workpiece integrity, but also economics, based on the time required to achieve the desired size and finish. For many years, probably due to the awkwardness of adjusting manual grinders, there were rules of thumb which suggested that the workpiece traverse rate, for roughing, should be between one half to two thirds the width of the grinding wheel for each revolution of the workpiece (see Fig. 7.7). A finishing traverse rate should be approximately one quarter the width of the grinding wheel per workpiece revolution and should never exceed half the grinding wheel width. For very high surface finishes the rate should be cut to 3 mm (0.125 in) per part revolution or less. These very basic rules of thumb appear somewhat arbitrary.

A more defined set of feeds and speeds are available and are best used under CNC conditions as the consistency of the feeds and speeds

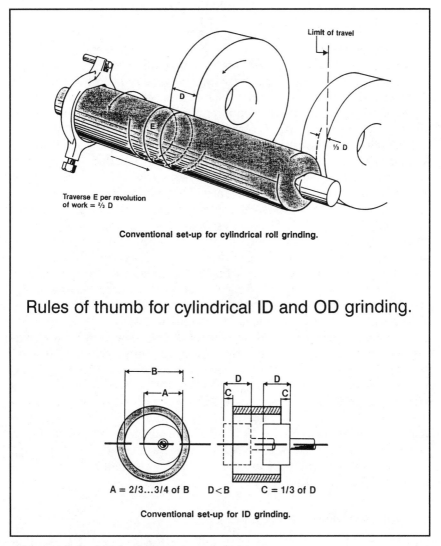

Figure 7.7 Cylindrical grinding.

will dramatically reduce the cut time. Depending on the material being ground, the work rotational speed should be set as follows:

The work speeds are shown in millimeters per second; feet per minute are in parentheses.

Type of workpiece material	Rough grinding	Finish grinding
Soft steels	150–250 (30–50)	250–350 (50–70)
Hardened steels	350–450 (70–90)	450–600 (90–120)
Cast iron	500–800 (100–160)	700–1000 (140–200)
Copper alloys	500–600 (100–120)	
Aluminum	750–1000 (150–200)	

The grinding wheel is typically going to run at 25 to 35 ms^{-1} (5000 to 7500 sfm) for conventional abrasives, a maximum of 42 ms^{-1} (9000 sfm) for diamond and faster than 60 ms^{-1} (10,000 sfm) for CBN. The grinding wheel traverse rate along the cylinder has to be such that the pitch of the grinding wheel is equally divisible into the width of the grinding wheel, e.g., for a grinding wheel 50 mm wide (2 in) the pitch can be 50 mm (2 in), 1 wheel width, 25 mm (1 in), 2 wheel widths, 16.667 mm (0.6667 in), 3 wheel widths, 12.5 mm (0.500 in), 4 wheel widths and so on. If a traverse rate is chosen which is not equally divisible into the wheel width, then feed-lines will occur along the cylinder due to the inherent, uneven wear across the face of the grinding wheel.

The normal, manual practice of preparing a cylindrical grinding wheel for plain, cylindrical grinding is to dress the wheel flat. So that fragmented grains do not compound the feed-line problem, the operator normally touches the edges of the wheel with a "Norbide" stick and breaks the sharp edge on the wheel. Depending on the direction of wheel travel during the grinding operation, the leading edge of the grinding wheel experiences a very high force and wears both unevenly and rapidly, reinforcing the feed-line problem.

It has been researched and found that if a shallow angle of approximately 1 degree is dressed into the face of the wheel, which is at least as deep as the wheel depth of cut (see Fig. 7.8), then the grinding forces are distributed more evenly across the face of the grinding wheel and a higher rate of stock removal can be achieved without the risk of workpiece surface damage. Where the combination of depth of cut and grinding wheel pitch along the cylinder come together there is an opportunity to use multigrade grinding wheels (see Fig. 7.9). However, this is usually an expensive alternative to simply dressing an angle on the sides of the grinding wheel.

7.3 Plunge Cylindrical Grinding

Plunge grinding is a cylindrical grinding process where the grinding wheel plunges into the workpiece at right angles to the axis of the part and forms a shape on the cylinder. Long and slender parts generally require special handling and fixturing, as the normal forces exerted on the workpiece during plunge form grinding are extremely high, high enough to deflect the workpiece and, in the extreme, deflect the machine tool. Therefore, it is necessary to use steady-rests to support the workpiece against the machining forces in order to minimize part deflection.

The center type cylindrical grinder is designed with four basic movements: the rotation of the grinding wheel, the rotation of the workpiece, the longitudinal traverse of the workpiece across the grinding wheel, and the infeed of the grinding wheel in either plunge or intermittent feed. These machines are not considered high-stock removal rate machines; they are precision finishing machines. However, as improvements are made in their design to improve rigidity and dynamic stability, their capability to remove large volumes of stock are creating new opportunities. One plunge grinding operation may take the place of a rough turning and finish grinding operation. A good example is the grinding of very delicate seal grooves (see Fig. 7.10). Turning such delicate seals generally causes distortion and very poor surface integrity, so the surfaces are generally finish ground. Plunge grinding can achieve both the full form seal in one plunge with no distortion, as the grinding forces are radial, and a superior surface integrity, which is generally associated with precision grinding.

Steady-rests are essential, especially when plunge grinding, or grinding flexible or delicate workpieces. The steady-rest performs the function of supporting the workpiece while it is being ground.

Steady-rests are used to support workpieces which have insufficient rigidity to produce satisfactory results in longitudinal, plunge, or internal grinding. Part deflection, due to the mass of the workpiece or the grinding forces, is counteracted by the support from the steady-rest. A further function of steady-rests is to prevent workpiece vibration and so eliminate chatter. There are four types of steady-rests (see Fig. 7.11):

1. Open fixed steady-rests with manual adjustment.
   The open steady-rest has two jaws and is manually adjusted. It does not allow grinding in the position of the steady-rest, as the size diameter will change and require adjustment in-process. This type of steady-rest is used on a machined diameter which is not ground. The two-jaw type has manual adjustment from a single knob or screw which moves both jaws in unison to a diameter. Another type of open, fixed, and manually adjusted steady-rest has a third support opposite the bottom bearing which secures the workpiece in a manner which prevents climbing of the back bearing. Polyamide materials are generally used for the steady-rest bearings. These materials wear rapidly, however, they leave little evidence of their

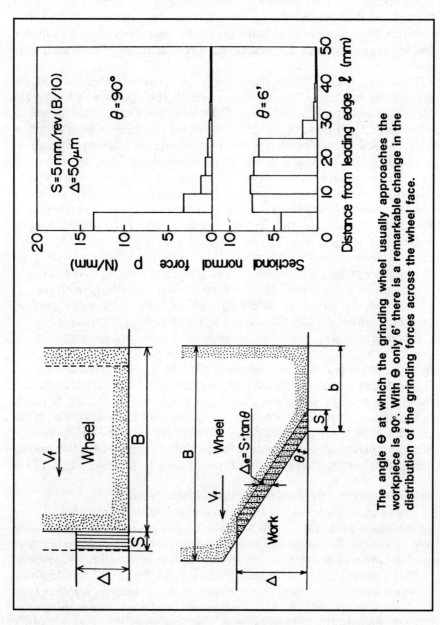

Figure 7.8 Distribution of grinding forces across a grinding wheel during cylindrical grinding. (*From work by T. Nakajima, K. Okamura, and Y. Uno at the Universities of Kyoto and Okayama in Japan.*)

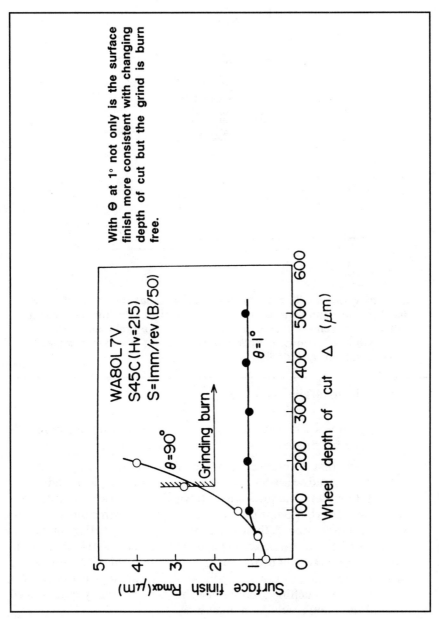

Figure 7.8 (Continued)

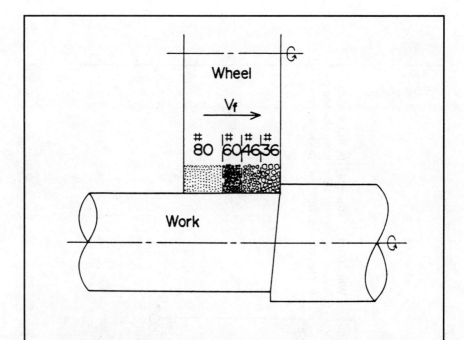

Figure 7.9 Multi-grade grinding wheel.

seating on the workpiece surface. These steady-rests are predominantly used for single and small batch production.

2. Open steady-rest with auto diameter adjustment.
   This type of steady-rest has two jaws which automatically adjust to the part diameter. The jaws travel towards the workpiece until contact is made; at that point the jaws are clamped with collets and can no longer be adjusted. Adjustment can only be made after retraction of the jaws. The workpiece may be loaded without restriction, as the jaws are retracted when the workpiece is changed. Grinding is not normally carried out in the support area. The steady-rest is usually seated on a premachined diameter. This type of steady-rest is used for medium batch production machines.

3. Open steady-rest with diameter dependent follow-up.
   This steady-rest operates automatically. It is diameter dependent and utilizes an in-process gage to feed back the part diameter to

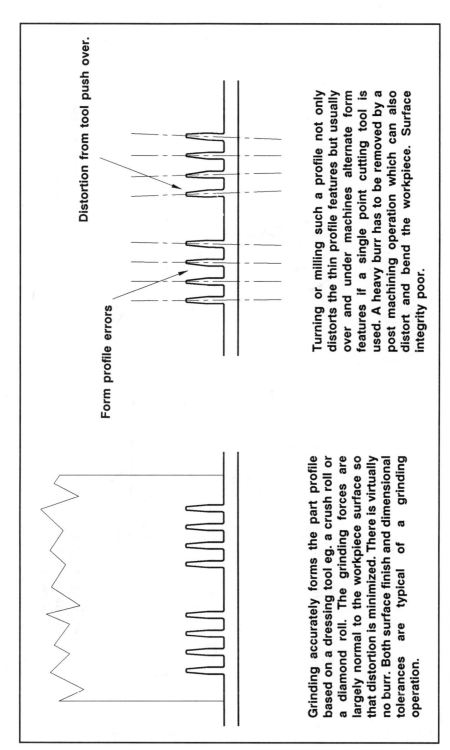

Figure 7.10 Plunge grinding.

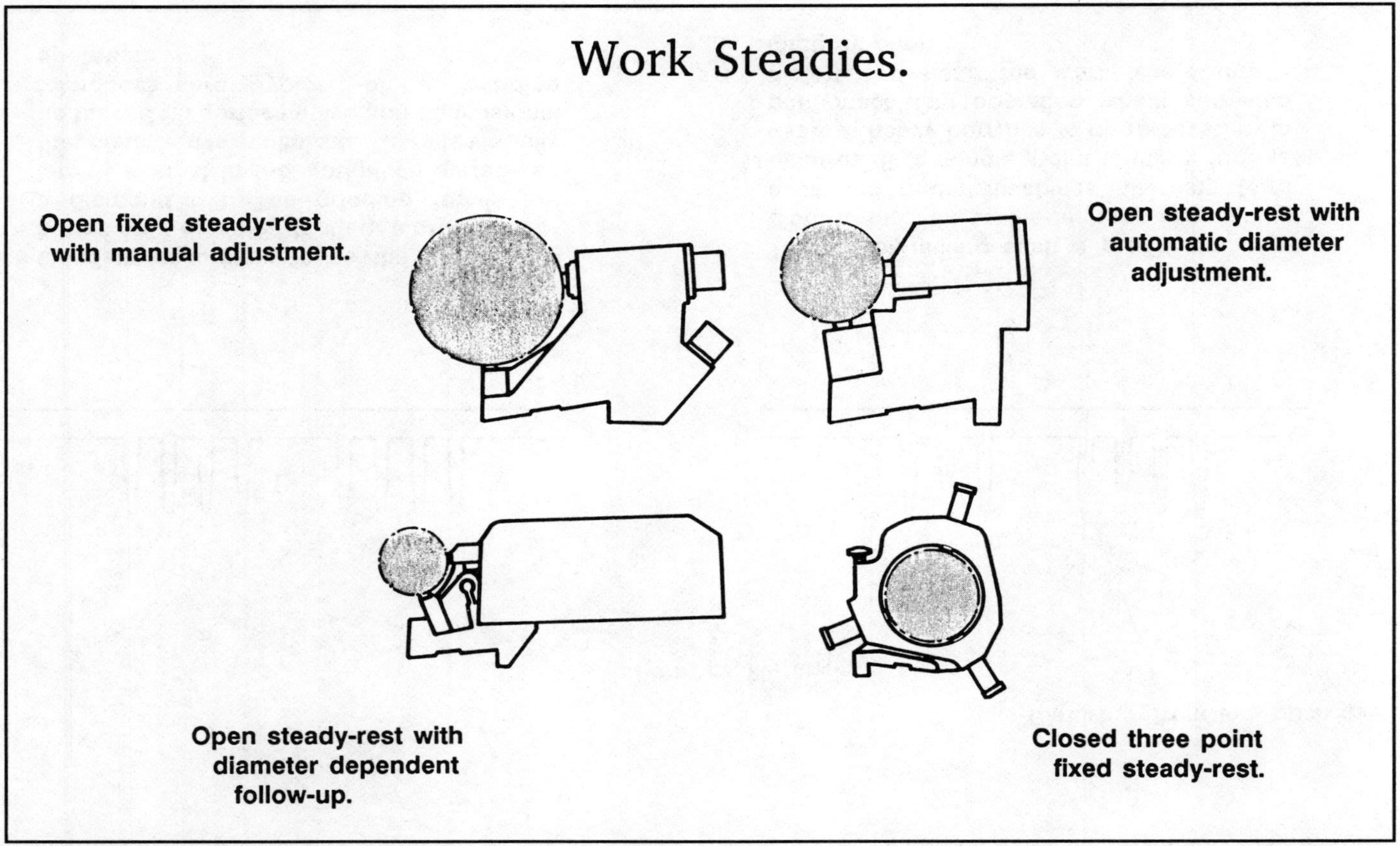

Work Steadies.

Open fixed steady-rest with manual adjustment.

Open steady-rest with automatic diameter adjustment.

Open steady-rest with diameter dependent follow-up.

Closed three point fixed steady-rest.

Figure 7.11 Steady-rests.

adjust for the jaw position. Alternatively, CNC may be utilized to achieve the follow-up. This type is used in conjunction with an automatic taper correction device and is used for high batch production.

4. Closed fixed steady-rest.
A closed fixed steady-rest is used on a premachined surface. It gives optimum support and achieves the finest accuracy. The workpiece is captive and allows plunge grinding of long workpieces. The very rigid support provides the rigidity to suppress self-excited vibration. This type of steady-rest does not allow grinding in the position of the steady-rest. Setup time for a closed fixed steady-rest may be lengthy. This type of steady-rest is used for high precision grinding or very long workpieces where more than one steady-rest might be used along the length of the workpiece.

7.4 Angle Approach Grinding

Situations often arise where grinding of a side face requires both a fine finish and a precise angle, face to diameter, relationship. It is not good practice to grind with the side of a parallel grinding wheel unless the side face of the grinding wheel is relieved (see Fig. 7.12). Therefore, angle approach grinding is generally used where the wheelhead of the machine is angled and dressed to grind the correct shape of two or more angled faces. The grinding wheel is plunged at an angle into the workpiece, achieving the final size on both the face and diameter with precise correlation of those surfaces. The process is very fast and accurate.

There are many advantages of angle approach grinding. Two or more surfaces may be completed more efficiently and in one plunge. The grinding marks on the face of the shoulder will also be concentric. This is desirable because seal faces with grinding marks, which are anything but concentric, aggravate a seal and cause heavy wear. Because the grinding wheel is grinding with its periphery, there is no side grinding/rubbing and, therefore, less chance of burning on the face. The large area of contact between the face and the diameter also tends to dampen any regenerative chatter which might occur. Angle approach grinding is not a suitable process when the shaft being ground is a splined shaft or has a key-way to interrupt the cut. The splines, during a combination angle approach grind on a face and splined shaft, will induce a pulse into the system, which will transfer into the surface of the face grind as noticeable radial marks. Such a grind is best performed in two separate grinding operations, first the face and then the splined shaft.

A common problem in cylindrical grinding is the occurrence of regenerative chatter, due to either the lack of rigidity of the machine system

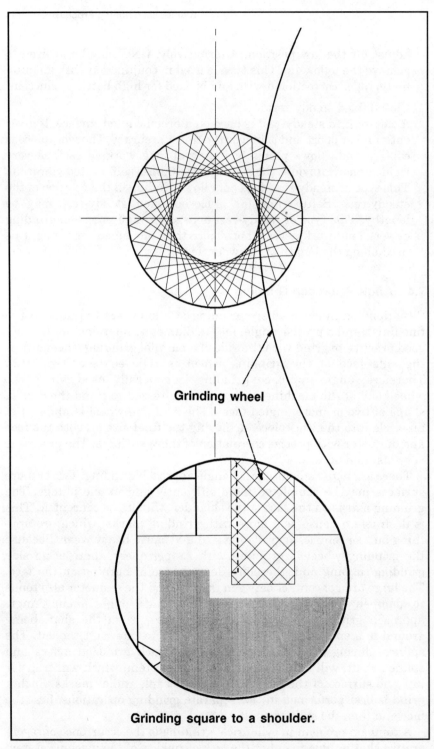

Figure 7.12 Two types of grinding marks on a ground face.

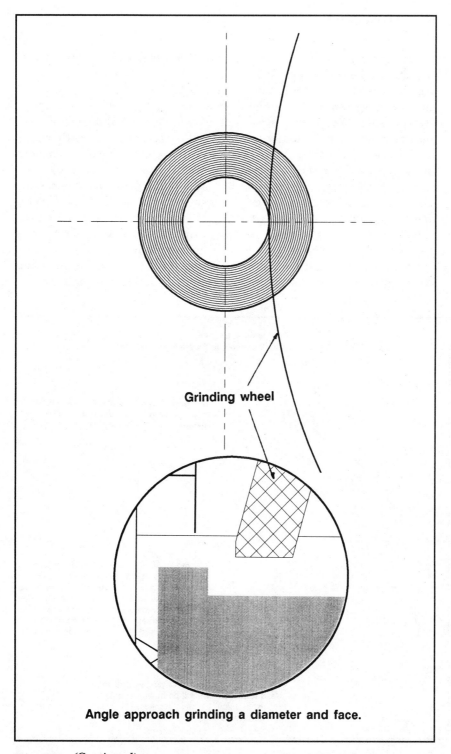

Figure 7.12 *(Continued)*

or the lack of rigidity of the workpiece. Regenerative chatter is a vibration due to the cyclic variation in the grinding wheel depth of cut, initiated by a machine vibration which occurs at a natural frequency of the machining system (see Fig. 7.13). Once the grinding wheel/workpiece system begins to vibrate at a resonant frequency, the grinding wheel and workpiece will cycle radially towards and away from one another. The motion creates a wave pattern around the workpiece which corresponds to the frequency of the vibration. The wave pattern on the workpiece, caused by the original vibration, produces cyclic forces on the system. The cyclic forces occur at the critical frequency initiated by the varying depth of cut of the grinding wheel, as it encounters the crests and the valleys of the wave pattern. As the natural frequency of vibration is regenerated by the crests and valleys, its amplitude increases and the machine can be heard and, in some severe cases, seen to vibrate. Once regenerative chatter has begun, it is extremely difficult to eliminate.

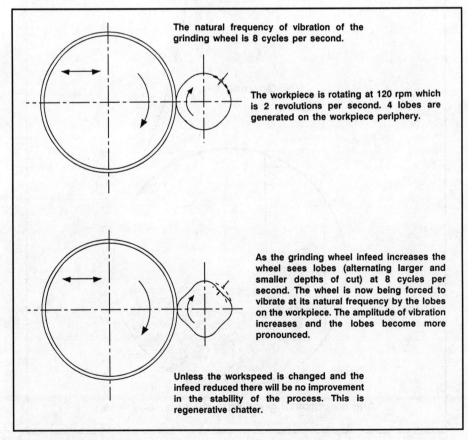

Figure 7.13 Regenerative chatter.

7.5 Internal Diameter (ID) Grinding

The internal grinding machine has the difficult task of the grinding wheel having to go inside the bore of the workpiece. The wheel is generally very small and therefore wears very quickly. Superabrasive grinding wheels for ID grinding are being used in increasing numbers due to their extended life between dressing operations. ID grinding wheels should be chosen carefully, not simply by the grade, but also by the dimension. The wheel diameter should be 70 to 80 percent of the diameter of the bore being ground. This will allow sufficient room in the bore for cutting fluid penetration and for chip clearance. The wheel width should be no greater than the diameter of the bore being ground. Larger diameter and width in wheels will block the way for the cutting fluid and load with grinding swarf. Also, because of the large area of contact, the forces on the spindle quill will be larger and may cause bell-mouthing of the bore. It is important to select the correct grade of wheel, nominally 80 grain and finer, as well as the correct peripheral speed of the wheel and workpiece. ID grinding spindles are high RPM spindles, and may be as high as 200,000 RPM in order to reach the desired cutting speed for the abrasive. The workpiece should rotate in the opposite direction to the grinding wheel at 150 to 250 sfm in the case of general internal grinding.

Because ID wheels have relatively small diameters, the ID spindles are similarly small. These quill shafts, in effect, involve a cantilever arm support and can deflect significantly against the grinding forces (see Fig. 7.14). It is also important that these spindles are carefully machined to high levels of concentricity, straightness, and run out, as the whirling forces on such a system may cause very poor grinding conditions. The amount of abrasive on an ID wheel usually means frequent dressing. CBN and diamond wheels in crush dressable and vitrified form have become popular for ID grinding, due to the long wheel life and the relatively low wheel cost with respect to the wheel size.

Clamping of the workpiece when ID grinding must be carried out very carefully. Due to the clamping forces acting against the grinding forces, it is typical, in the case of three and four jaw chucking, to encounter three and four lobed bores (see Fig. 7.15).

7.6 Cam Grinding

Cam grinding is a very special form of plunge grinding (see Fig. 7.16). It is a very complex and most accurate process. The cam to be ground is usually held between centers, or in a hydraulically operated chuck, with special steady-rests located on the bearing diameters of the cam shaft. It is the latter system which achieves the most accurate chucking of the cam. Chucking on the bearing diameters simulates the mounting of the cam shaft in the engine, and therefore guarantees

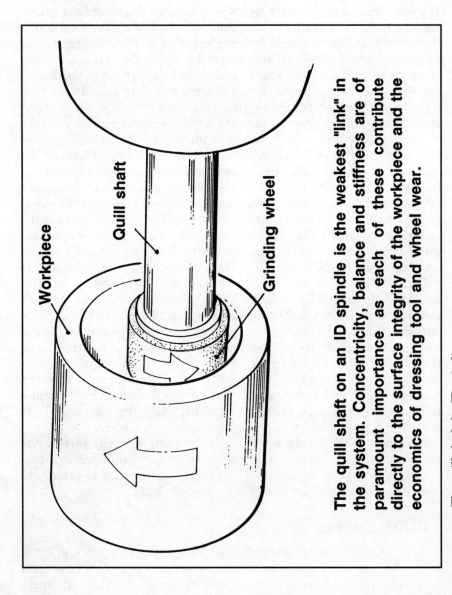

Figure 7.14 The quill shaft for ID grinding.

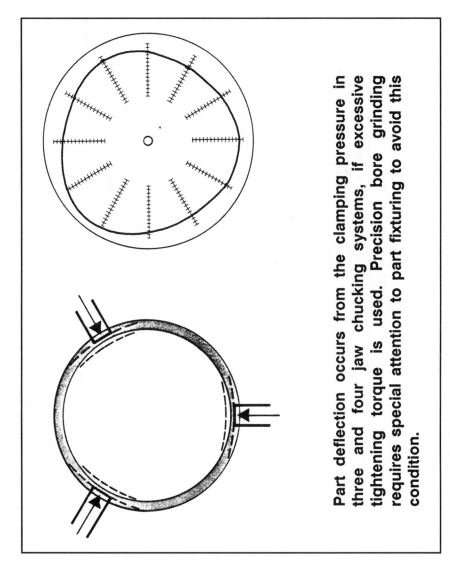

Figure 7.15 Clamping force.

Part deflection occurs from the clamping pressure in three and four jaw chucking systems, if excessive tightening torque is used. Precision bore grinding requires special attention to part fixturing to avoid this condition.

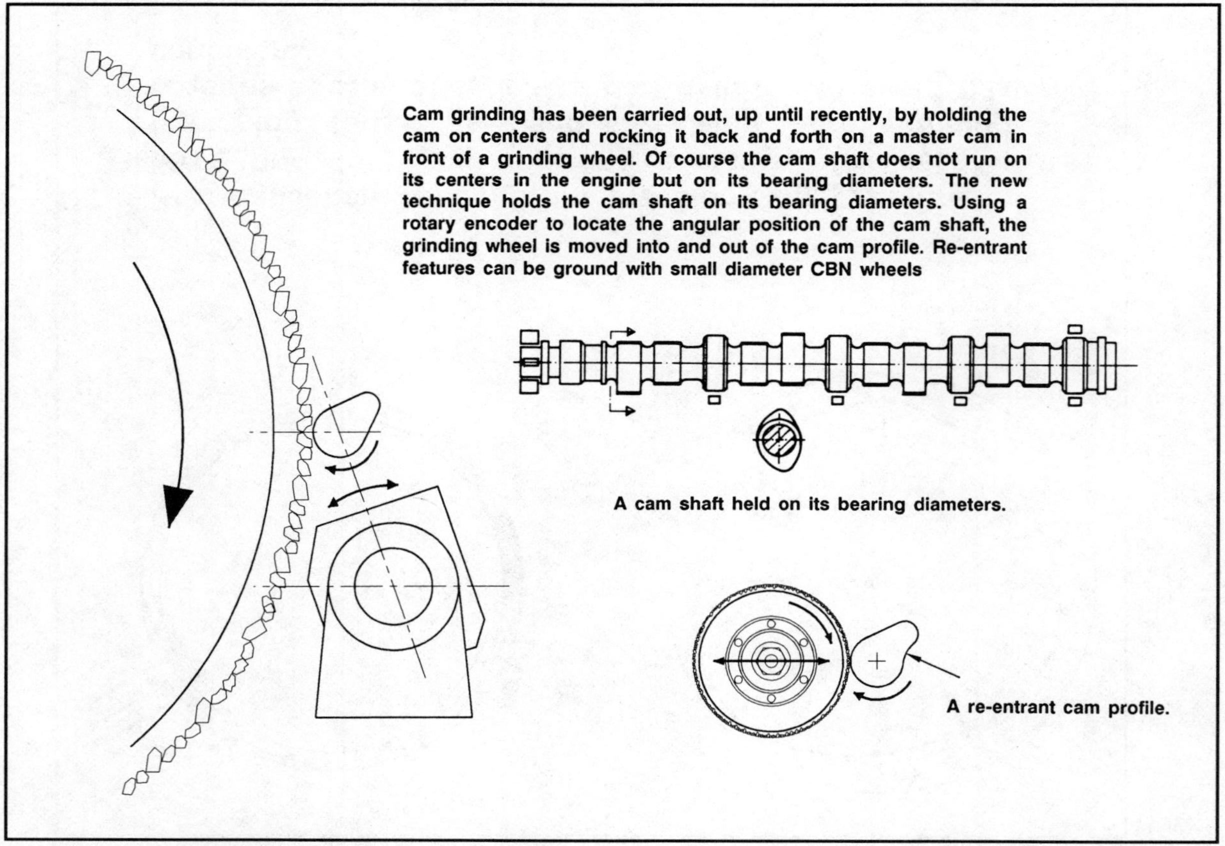

Cam grinding has been carried out, up until recently, by holding the cam on centers and rocking it back and forth on a master cam in front of a grinding wheel. Of course the cam shaft does not run on its centers in the engine but on its bearing diameters. The new technique holds the cam shaft on its bearing diameters. Using a rotary encoder to locate the angular position of the cam shaft, the grinding wheel is moved into and out of the cam profile. Re-entrant features can be ground with small diameter CBN wheels

A cam shaft held on its bearing diameters.

A re-entrant cam profile.

Figure 7.16 Cam grinding.

higher concentricity of the base circle in relation to the bearings. The workpiece is rigidly held and mounted on a torsionally stiff, rocking table. The rocking motion of the table creates the cam profile by moving the workpiece in and out of the grinding wheel. The rocking table is positioned by a series of master cams preset and spring-loaded onto a rotating drive. It is important that the grinding wheel maintains the desired velocity profile relative to the angle of rotation of the cam. As the cam follower, in the engine, changes velocity, the grinding wheel must effectively take the same path, although it is the cam shaft which is moving during the machining operation. In a modern cam grinder, a microprocessor receives a signal relative to the angular position of the cam as it is rotating and determines the correct velocity profile around the cam being machined. The sensor feedback to the microprocessor control not only improves the accuracy and quality of the cam, but also significantly reduces the grinding time, particularly in the spark-out phase. It is important to control the stock removal rate of the grinding wheel. As the rough cam rotates, the arc of contact will vary with the angular position of the cam with respect to the grinding wheel. If the RPM of the cam shaft were kept constant, the lobed shape of the cam would significantly change the stock removal rate as the cam shaft rotated. On a high-velocity rising section, the effective depth of cut of the grinding wheel may cause thermal damage, softening the cam surface due to excessive heat input from the overload condition. Cam grinding is a complex process to control and although, historically, success has been enjoyed with semimanual systems, the use of a superabrasive CNC cam grinding machine is by far the best method, achieving the most accurate and consistent cams.

Cam grinding in the automotive industry is one of the most complex processes, yet it enjoys the benefits of high automation. Cam shafts are typically machined on a transfer line, which takes the rough forged or cast cam-shaft and not only completes the machining of the rotary features, cams, and bearing diameters, but also inspects and records the dimensions of each workpiece automatically.

New and more efficient internal combustion engines are being designed with a reentrant cam profile. This is a profile which has concave features. Such cam profiles require the use of much smaller grinding wheel diameters to accommodate the reentrant feature and are spawning a new era in cam grinders, which have been specifically designed to use small diameter superabrasive wheels.

7.7 In-process Gaging

Cylindrical grinding is a very precise and highly productive machining process. Large batch production of parts on a cylindrical grinding

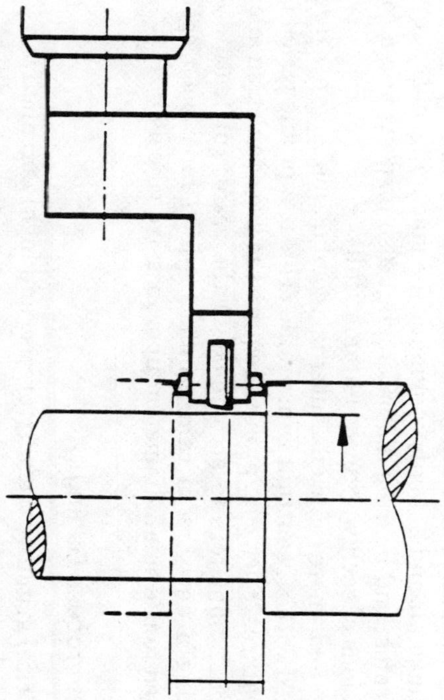

The in-process gage head is a touch type probe which can measure both the part diameter and the position of the flange face. The gage can be used initially to locate the part's position prior to grinding to offset the CNC control. The gage, with its very hard carbide or sapphire styli, is then used to measure the part to confirm its dimensions during and after grinding.

Figure 7.17 In-process gaging.

machine requires in-process gaging. Across a series of manual operations, there can be a great deal of movement of the part in and out of the machine for measurement. Each relocation of the part increases error, and as final size is approached, the grinding becomes ever more tedious as the operator attempts to achieve the required surface finish and the exact diameter at the same time.

In-process gaging allows the part to be very accurately measured while it remains in the machine. Modern cylindrical grinding machines have the in-process gage tied into the CNC control system, so that the entire grinding cycle is automated. The machine effectively inspects itself and the parts are consistently ground to size.

The in-process gaging system operates by two styli touching the workpiece periphery. The position of the styli are fed back to the control system, which then positions the grinding wheel with respect to the diameter of the workpiece. Such systems can also very accurately measure interrupted diameters, like splined shafts, by an adjustment of the damping in the styli fingers. Further sophistication is the ability to measure to a shoulder, thus combining the measurement of the diameter with its position along the shaft (see Fig. 7.17). A CNC grinding machine, in combination with "Flag" gaging and in-process measurement, can locate a part's position, perform a series of diameter and step grinds relative to that flag location, inspect all aspects of diameter and land widths within very close tolerances, consistently and fully automatically.

Chapter 8

Flat Surface Grinding Processes

The grinding of any surface, no matter what shape or form, may be termed surface grinding. In abrasive machining, the term surface grinding is generally given to the production of a flat, angular, or contoured surface, which is created by feeding the workpiece in a horizontal plane beneath a rotating grinding wheel. A profile may be dressed onto the grinding wheel periphery using a crush roll or a diamond roll. A profile may also be traced onto the periphery of a grinding wheel by using a mechanical, pantograph-type tracing device attached to a single point diamond. CNC may be used to trace a very accurate profile across a grinding wheel periphery, using either a single point diamond or a rotating diamond-impregnated disk. The diamond disk is somewhat akin to a very thin diamond roller. When grinding takes place with anything other than a flat profile on the grinding wheel, the process is generally called profile or form grinding.

8.1 Reciprocating Grinding

A reciprocating grinding machine (see Fig. 8.1) has a horizontal spindle and a table, which, dependent on the machine size, may vary from 300 mm (12 in) on up to 4.5 m (15 ft) or more in length, with a proportional width of table ranging from 75 to 100 mm (3 to 4 in) up to 1.5 m (5 ft) or more. The workpiece is mounted on the table, which moves back and forth beneath the grinding wheel. The table might be hand actuated on very small machines. On larger machines the table is usually driven hydraulically or driven by a ball-screw and actuated by a DC-motor. There are also machines built which drive the table on highly tensioned toothed belts.

To achieve the cutting action, the grinding wheel is fed down into the

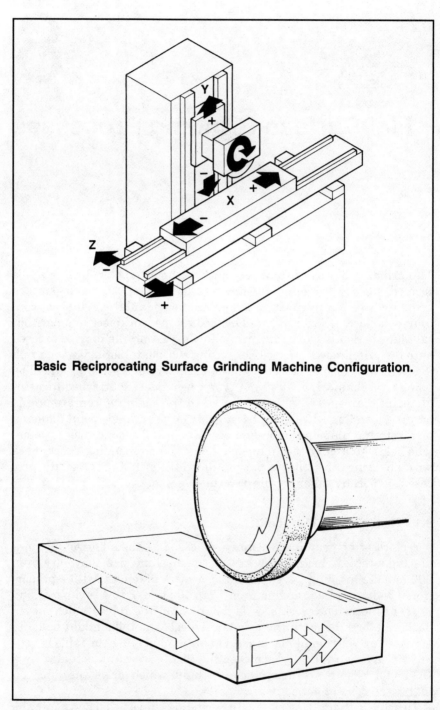

Figure 8.1 Reciprocating grinding.

workpiece. This is the infeed motion. Older machines were designed where the grinding wheel was fed by hand. Later machines were linked to an infeed mechanism, which fed the grinding wheel down by a set amount of down-feed that occurred at the end of each table stroke or cross-feed stroke. Modern machines, with electronic controls, are able to be set in a programmed cycle to infeed precise amounts per stroke, using stepper motors, DC servo motors, etc. Automated machines may even have automatic size capability, where the position of the wheelhead is sensed and fed back to the control system so that the part being machined is machined accurately, without manual intervention to measure the workpiece. More accurate systems have in-process gaging interfaced with the machine control. The in-process gage monitors the size of the actual piece being machined, as it is machined, so that very accurate size control may be maintained.

It is common for most reciprocating surface grinders to have a cross feed. Cross feed is a machine motion which is made at right angles to the table movement. Cross feed is made automatically in conjunction with the table feed to allow the grinding of both long and wide parts. The cross feed is not a cutting axis. These movements are better illustrated by taking a closer inspection of the construction of a reciprocating grinding machine.

The machine saddle is mounted on the base guideways. The saddle may be moved precisely towards or away from the column of the machine. The machine base guideways may also be termed the cross-slide guideways. The saddle has very precise and well lubricated guideways machined at right angles to the cross-feed. The ways in the saddle are the table guideways and are similarly well-lubricated, so as to avoid any stick-slip. The table may be manually moved or automatically moved using hydraulic or electrical feed motions. Both the cross feed and the table feed are very precise movements with a very high accuracy of parallelism and flatness. The end of each table travel movement is determined by trip-dogs or microswitches, which actuate the reversal of the table movement. Depending on the length of the workpiece, a second set of trip-dogs or microswitches may be set to accommodate different lengths of workpiece, shortening the stroke of the machine in order to economize on the air-cut time. Modern CNC surface grinding machines have the travel stops preprogramed in software. End stops, even on CNC machines, tend to be more basic hard stops to avoid the machine overrunning its available travel.

A most important part of any machine tool system is the construction of the guideways. The qualities which constitute a superior machine are rigidity, accuracy, freedom from stick-slip, good damping, and longevity. A very precise guideway system is that of pre-loaded roller guideways (see Fig. 8.2). The guideways are constructed from plain,

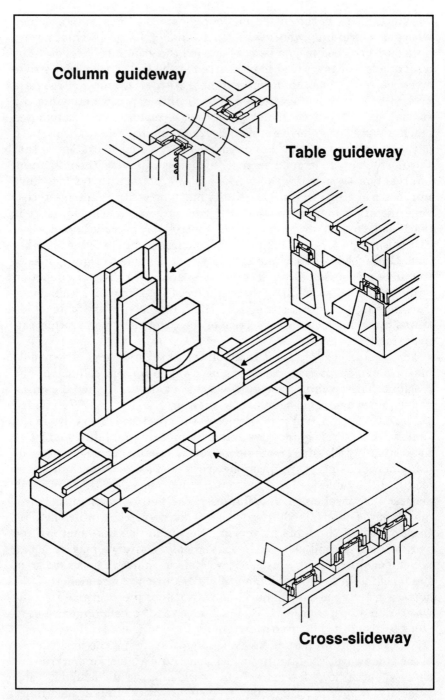

Figure 8.2 Pre-loaded roller guideway bearing system.

hardened steel ways, accurately machined as linear bearing surfaces for small needle roller bearings. Once assembled, the guideways are pre-loaded so that a very accurate, rigid, and precise guideway system is created. The roller guideway system is a very low friction system, and may be used for very precise machining, using encoder feedback from a stepper motor drive. One drawback is the propensity for these guideways to transmit process vibration throughout the machine system. However, pre-loaded roller bearings may be used in combination with a machine feature which helps to dampen the vibration, e.g., an epoxy granite base or a weldment base and column.

Many machines have been designed with the traditional V-flat ways. Such systems, which have more frictional drag, may require precision glass scales to verify the slideway position. The V-flat ways require a film of lubrication oil to be present between the mating surfaces. This film may cause inaccuracies, due to compression of that film under load. At various table speeds the thickness of the oil film may increase due to the hydrodynamic effect of the table riding on the oil film. Still the oil film provides a good damping coefficient. Dependent on the relative size of the V-flat system, it may have properties comparable to the pre-loaded roller guideway design. Generally V-flat ways have superior vibrational damping (see Fig. 8.3).

Hydrostatic bearings are yet another alternative. Though hydrostatic bearings are extremely rigid bearing systems, they may suffer from thermal growth due to the varying temperature of the hydraulic oil. Moreover, hydrostatic systems require the strictest discipline in maintenance and cleanliness of the hydraulic oil and pumping unit (see Fig. 8.4).

Though there are many sizes of reciprocating grinding machines, the constituent parts of their construction are basically similar. The machine base is usually fabricated from cast iron, which provides a large mass for static stability of the machine tool. Some older machines will use the hollow structure of the casting to house the hydraulic reservoir and pumping system for the table drive. This was quickly found to be detrimental to the accuracy of the machine tool, due to thermal instability. As a move to improve accuracy and repeatability, machines have been built where the cutting fluid, which is refrigerated, is passed through the machine base casting and spindle housing to provide thermal stability to the structure. The hydraulic reservoir and pumping system on a modern machine is typically a self-contained unit, separate from the machine tool base. The need for both static and dynamic vibrational stability has prompted some machine tool builders to construct their bases from an epoxy concrete which has superior thermal stability and vibrational damping properties. In fact, not only grinding machines have been built from epoxy concrete (see Fig. 8.5), but also lathes and milling machines.

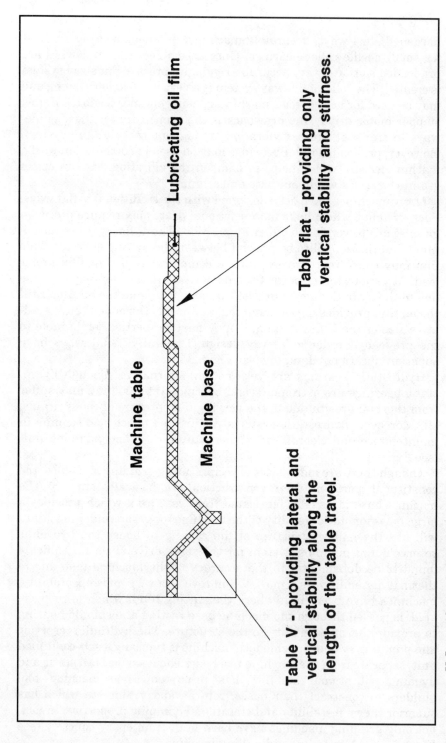

Figure 8.3 V-flat guideway bearing system.

The column of a grinding machine has vertical ways; these ways are the guideways for the wheelhead assembly. The wheelhead assembly contains the grinding wheel spindle, mounted in precision bearings and located in the spindle bearing housing. The grinding wheel is held between flanges, which fit onto the spindle nose. The grinding wheel spindle bearings are loaded in order to take up any end thrust which might occur in the grinding process. The spindle runs true and in a plane parallel to the machine table. The infeed mechanism on the wheel head is typically a leadscrew or precision ball-screw. The infeed mechanism is able to feed accurately in increments of 0.001 mm (0.00005 in) and less, depending on the type of stepper motor and/or feedback system. Both the wheel head and the cross slide usually have a rapid traverse separate from their incremental feed capabilities. The accuracy of these slideways, particularly for manual operation, usually depends on glass scale calibration for exact positioning of the wheel to the workpiece.

The grinding spindle is a cantilever spindle, which means it has no outer support. The highest forces are generated through the grinding wheel on the end of the spindle. Hence, the construction of the spindle bearings, spindle nose, and vertical slideway are critical to achieve accuracy and minimize vibration and flexibility under load.

From the fundamental mathematical model and referring to the micro-milling analogy in Chap. 4, it was appreciated that the grain depth of cut is very small. It follows that the amplitude of vibration of a grinding machine system need not be very large to significantly affect the satisfactory operation of the grinding process (see Fig. 8.6). The amplitude of vibration will cause active grains on the grinding wheel periphery to become inactive as the wheel rises and then become overloaded as the wheel falls, increasing and decreasing the grinding wheel depth of cut. There is generally sufficient damping in the grinding system to minimize process vibration. However, it is most important to insure that the grinding process is operating properly. The grinding wheel must be dressed correctly and the feeds and speeds of the grinding wheel and workpiece must be determined and monitored throughout the machining cycle, so that process vibration is kept to a minimum.

There are many machines on the market with sufficiently automated cycles, CNC, and feedback mechanisms to eliminate a great deal of the repetitive size checking of the part. Taking the part out of its machining position to check for size and then attempting to place the part back in exactly the same place is very difficult and, in most cases, leads to the major source of error.

The small infeed depth of cut and the fast feed rate associated with reciprocating grinding inherently cause the grinding wheel to glaze. Glazing is a term given to the attritious wear on an abrasive grain due

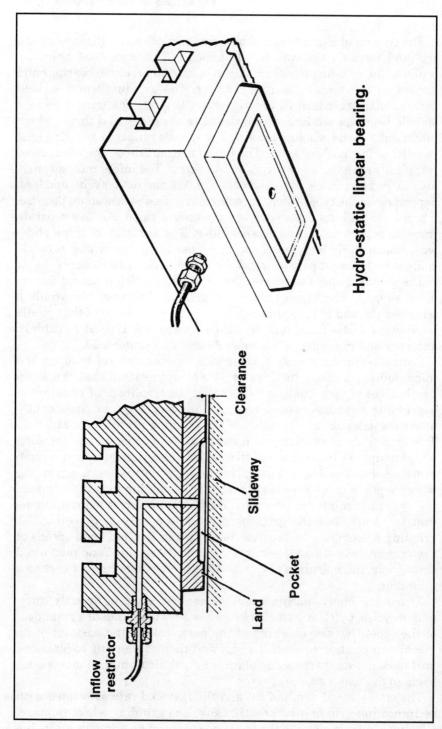

Figure 8.4 Hydrostatic bearings.

Hydro-static journal bearing.

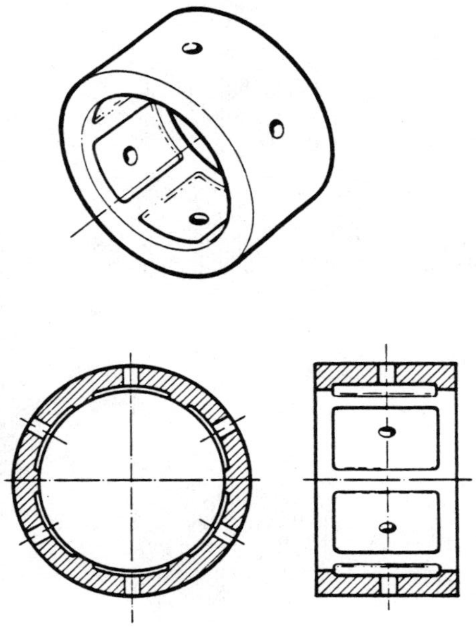

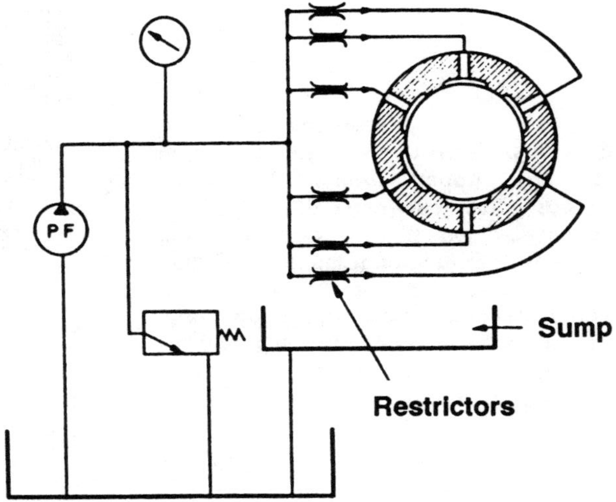

Hydraulic circuit for a hydro-static bearing.

Sump

Restrictors

Figure 8.4 *(Continued)*

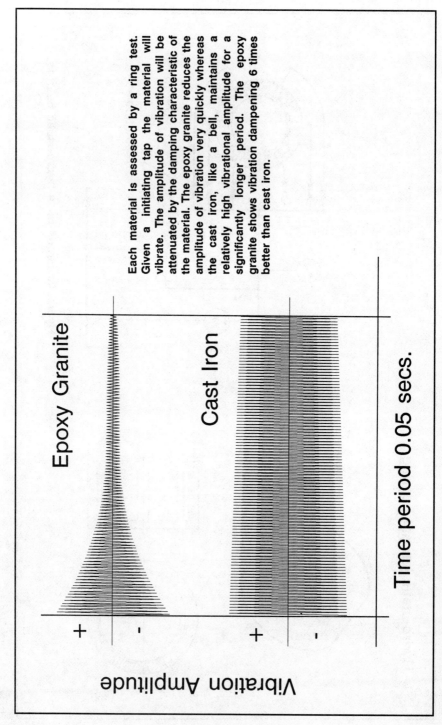

Figure 8.5 The vibrational stability of epoxy granite and cast iron.

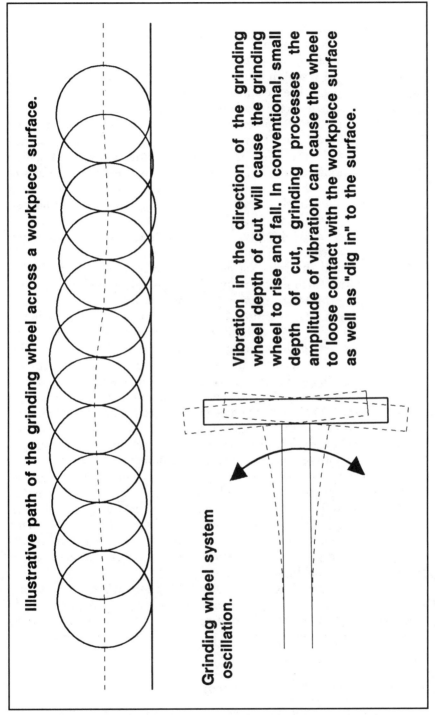

Figure 8.6 Vibrational instability of a grinding wheel.

to excessive rubbing. Machining with a grinding wheel in a glazed condition may result in thermal damage to the workpiece surface and induce tensile stresses, which could lead to cracking. The dull nature of a glazed grinding wheel in the reciprocating grinding process exerts a high normal force on the workpiece and opposing forces on the machine spindle. Hence the need for spark-out cycles. A spark-out cycle is where the machine is left to reciprocate across the workpiece after reaching final size on down-feed. The normal grinding forces build up on the way to final size, so that there is an inherent deflection in the machine tool once apparent size has been reached. Letting the machine reciprocate under a no-feed condition allows the elastic deformation to relax at close to zero. During the machining cycle, the forces are periodic on the spindle. Due to the traditional cantilever design of machine spindle, it will eventually initiate vibration and chatter, further degrading the surface integrity of the workpiece.

Having covered the grinding of flat workpieces on a reciprocating surface grinding machine, there are other methods for achieving a satisfactory flat surface that are worthy of mention.

A vertical spindle machine (see Fig. 8.7), commonly called a Blanchard grinding machine and manufactured by Cone-Blanchard, has become a very popular design for the high-stock removal of flat stock. The machine does not use a grinding wheel, but segments of abrasive. The segments are assembled into a shape resembling a recessed grinding wheel. The "wheel" is mounted on a vertical spindle above a rotary table. The table axis is offset from that of the grinding wheel, and so allows the loading of parts without raising the wheel. The wheel and rotary table turn in opposite directions and are brought towards one another. The infeed rate on a Blanchard machine can be very high and can remove large amounts of stock very quickly. Because of the rotary nature of the wheel and table, it is relatively easy to achieve a flat and parallel surface on a workpiece, though the pattern of the grinding marks are a complex intersection of circle arcs. This surface is particularly favored by fast-food restaurants. The intersecting arcs provide just the right amount of grease retention on a hot plate surface for cooking hamburgers.

Rotary surface grinding is a process waning somewhat in popularity. The rotary grinder (see Fig. 8.8) configuration grinds with a pattern of concentric rings, ideal for the mating surface of flanges. The rotary grinder is also capable of grinding slightly concave or convex lens shapes.

8.2 Form Surface Grinding Processes

Form grinding is a surface grinding process which machines an accurate shape, generally parallel to the axis of the table traverse. An

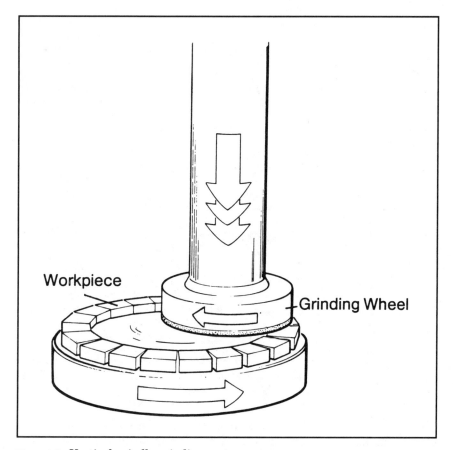

Figure 8.7 Vertical spindle grinding system.

example might be a form ground on the edge of a carbide insert. The form would be dressed by crushing, diamond roll dressing, CNC diamond disk dressing, or pantograph dressing, using a single point or "fliese" tool. The machine would be operated in the same manner as for grinding a flat surface, except that no cross-feed is used.

Strict attention must be paid to the condition of the grinding wheel and the process parameters when form grinding very accurate work, particularly if the grinding wheel is allowed to become dull and the workpiece is very thin. Heat will be generated in the surface of the workpiece, which will either cause the workpiece to expand and distort during the grinding operation or create such high residual stresses that once the clamping on the part is released, it will spring into an arch. Particular care must be exercised when "sparking-out" on thin parts. Sparking-out not only relieves the grinding forces on the grinding wheel, but also may generate a great deal of heat, due to the scrub-

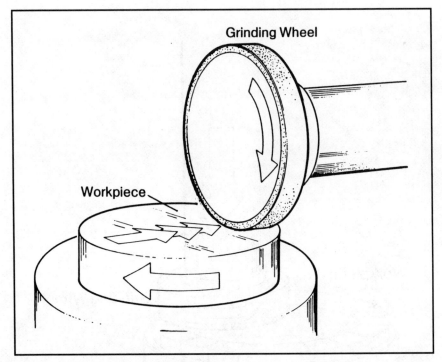

Figure 8.8 Rotary surface grinding system.

bing action of the cut. It is not unusual for a workpiece to be reasonably stable during the grinding, but on spark-out to distort up into the grinding wheel, due to the heat input, causing the material to warp. This is particularly so with superalloys and heat sensitive materials. Clamping methods may assist the situation. Clamping the part tightly may induce high stresses, due to the restraint of the part. A clamping method which may overcome this is a "free state." The workpiece is placed on a magnetic table and a set of blocks or keepers, made from steel, are placed around the workpiece. Once the table has been magnetized, the keepers will prevent the part from skidding, yet hold it in position for grinding. A sharp wheel with plenty of cutting fluid flow is essential at all times.

Surface grinding is a most precise process. It is extremely accurate and provides very fine surface finishes. The results, however, depend on the stiffness of the machine tool, the correct wheel selection, the dressing technique, and the correct method of clamping. Poor surface finish is sometimes a problem in surface grinding. Chatter may be the cause, due to a bad set up, an unstable machine, a poor grinding wheel, and/or a dressing technique. Surface blemishes can arise, such as slashes or flecks in an otherwise good finish. This may be due to poor filtration of the cutting fluid. Free abrasive grain is taken by the fluid between the grinding wheel and the workpiece, causing a deep scratch in the surface.

Depending on the material and the amount of stock to be removed, plunge form grinding may be substituted by creep-feed grinding. Particularly where the material is difficult to grind, the reciprocating form grinding process suffers from the inability to hold form accurately for a long period of time, unless the machine has the capacity to use superabrasive wheels. Each time the grinding wheel impacts the corner of the workpiece on each table stroke, part of the grinding wheel is broken away. Much in-process time is therefore taken up in redressing the form on the grinding wheel. The economics of the process must be assessed in the light of creep-feed grinding, taking into account batch quantities, material, stock removal, etc. Reciprocating plunge form grinding may be inefficient, not only from the aspect of frequent grinding wheel redressing due to glazing, loaded material, or the lack of form, but also from the air-cut time spent at the end of each stroke when the table has to stop, reverse, and accelerate to speed (see Fig. 8.9).

8.3 Creep-feed Grinding

Creep-feed grinding is a process which combines high-stock removal, precision size, form, and surface finish. There are many features of creep-feed grinding which are specific to the process and not of a generic nature; however, the basic principles of grinding still apply (see Fig. 8.10).

The construction of a creep-feed grinding machine is quite different from the conventional reciprocating grinder. Creep-feed grinding exerts much higher forces on the workpiece and requires tight control over the table feed. One basic difference is that the table drive should be a positive mechanical drive as opposed to the hydraulic drives typical for reciprocating machines. Hydraulic drives have been used for creep-feed grinding, but only in the up-cutting mode where the horizontal forces act against the hydraulic thrust force. Higher spindle horsepowers are necessary for creep-feed, and hence, the machines have to be constructed to handle the power and the forces expected in creep-feed grinding. Vibration is less of a problem for creep-feed grinding, as the area of contact between the grinding wheel and workpiece is much greater than in reciprocating grinding and tends to dampen any process vibration. This is not to say that vibration does not occur. Vibration in the creep-feed grinding process occurs from an out-of-balance wheel, stick-slip in the slideways or drive system, poor dressing, and machines that lack rigidity.

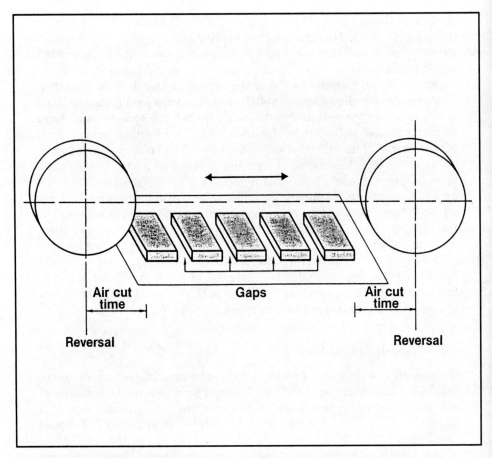

Figure 8.9 Reciprocating plunge form grinding.

Wheel speed (RPM) should be a variable in creep-feed grinding in order to keep the grinding wheel surface speed constant. The application of the cutting fluid is critical. A high volume of cutting fluid is applied to the process. It is advantageous if the speed of the cutting fluid from a jet nozzle equals or exceeds the peripheral speed of the grinding wheel in order to achieve penetration into the arc of cut. Typically, a cutting fluid system for creep-feed grinding has high pressure, up to 10 bar (150 psi), and high flow rate, 5 to 10 liters/s (80 to 160 U.S. gpm), pumps with large tanks to give adequate time for the grinding swarf to settle and filter. Refrigeration is advantageous in keeping the process in control, as the thermal aspects of the process are the most critical.

The creep-feed grinding process has many advantages over the reciprocating form grinding process. In particular, creep-feed grinding has a

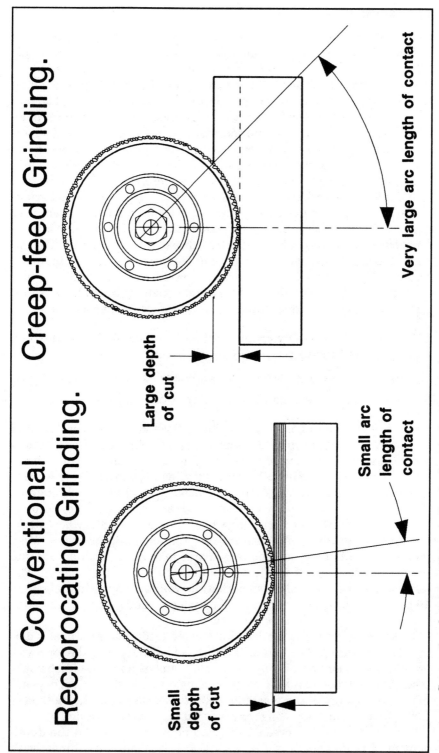

Figure 8.10 Conventional reciprocating grinding and creep-feed grinding.

higher capacity for the stock removal of "difficult-to-machine" materials and the ability to maintain close tolerance forms for longer periods of time. The process is analogous to climb milling, but with a grinding wheel in place of the milling cutter. Depths of cut in the order of 25 mm (1 in) are not uncommon in tool steel materials. Depths of cut are somewhat less for heat resistant and thermally sensitive nickel-based superalloys. The feed rate of the workpiece into the grinding wheel may be as low 12 to 25 mm/min (0.500 to 1 in/min). However the grinding wheel only contacts the workpiece once, and therefore maintains its dressed form for a longer period of time. The area of contact in creep-feed grinding is very large due to the very long arc of contact. The forces are very high on the workpiece, however, because they are spread over a large area, the force on the individual active abrasive grain is small. Though a great deal of stock may be removed in one pass, it is typical to take a second "clean-up" pass after a dressing cycle, to ensure flatness and accuracy of the dressed form. The overall benefit of the process is productivity with high surface integrity. A typical cycle runs as follows:

1. With the wheel freshly dressed, a final skim cut of 0.05 mm (0.002 in) is taken across a previously roughed part.
2. With the wheel in that condition, it roughs the next part.
3. The wheel is dressed and makes a final skim cut as in 1.

Creep-feed grinding is purely a plunge form grinding operation and has no cross-feed. The grinding wheels required for creep-feed grinding are very special and herald a new chapter in grinding wheel technology. The wheels are most commonly vitrified aluminum oxide wheels. They are very porous and quite soft, usually in grades D through G, though harder grades have been used successfully. The porosity induced into a creep-feed grinding wheel gives a very open wheel structure, which provides chip clearance for the long chip produced. Most important of all, however, the large pores take more cutting fluid into the arc of cut to assist in the dissipation of the grinding energy. The structure of these wheels is such that, by volume, approximately 30 to 40 percent is abrasive grain, 5 to 10 percent is bond material, and 50 to 60 percent is air-induced as porosity.

The preferred method for dressing a creep-feed grinding wheel is diamond roll dressing. Not only will the wheel be sharp, but the form will be controlled more accurately. For small batches and one-off jobs, it is costly to use diamond rolls. Crush dressing may be substituted, depending on the rigidity of the machine tool setup. Otherwise, the method of CNC, diamond disk dressing may be used to dress the profile.

The one or two pass creep-feed grinding process spurred the development of twin wheel grinders (see Fig. 8.11). These machines, with special tooling, can grind both sides of a workpiece at the same time. Symmetrical form accuracy may be achieved with zero mismatch if dressing from the same diamond roll can be achieved. A great benefit from twin wheel grinding is that the normal forces on the grinding wheel are equalized and have less of a resultant force on the workpiece.

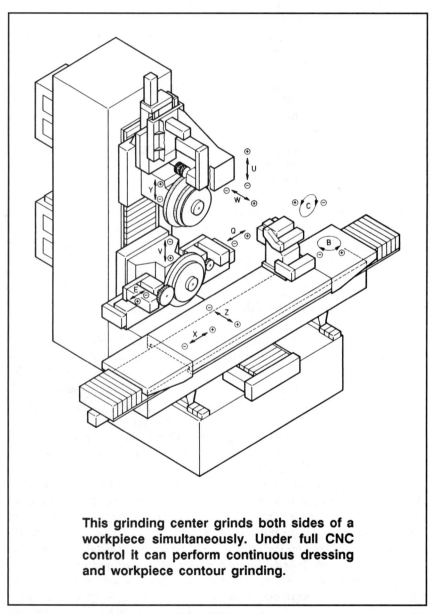

This grinding center grinds both sides of a workpiece simultaneously. Under full CNC control it can perform continuous dressing and workpiece contour grinding.

Figure 8.11 Twin spindle, multi-axis grinding center.

During the mid-1970s a great deal of research was carried out in the area of creep-feed grinding. It was discovered that by continuously dressing the grinding wheel with a diamond roll dresser and at the same time grinding the material, the specific energy of the process decreased dramatically; this allowed a phenomenal increase in the stock removal rate. Parts which previously took minutes to grind were easily produced in a matter of seconds. The added bonus of this method lies in the fact that there is a quantum leap in the productivity of the process; the risk of thermal damage to the workpiece is virtually eliminated.

All grinding wheels degrade after dressing, some faster than others. The grinding wheel is at its sharpest just after it is dressed. However, as soon as it touches the workpiece, it begins to get dull or load up with material. A continuously dressed grinding wheel stays in the peak of sharpness all the time, minimizing the rubbing energy and allowing incredible stock removal rates.

Continuous dress creep-feed grinding is now available on most creep-feed grinding machines. The machine controls required for such a system are more complex, in that the grinding wheel must be kept at a constant surface speed with respect to the diameter change, and the position of the wheelhead must be automatically compensated as the grinding wheel gets smaller throughout the grinding cycle. In addition to the above control, the ability to contour a profile is added. A dressed form may be ground into a workpiece with respect to a machine-controlled contour, rising and falling with the direction of the table traverse (see Fig. 8.12).

The continuous dress process led to a number of developments which have changed the nature of abrasive machining systems. Continuous dressing provides such a highly productive cutting time, that the aim of further and future work is to reduce the time for loading and unloading the workpiece, which manual labor would otherwise pace. Traditionally, the reciprocating grinding process spent most of the floor-to-floor time in the cutting cycle. Continuous dress creep-feed grinding has reversed the situation, such that most of the floor-to-floor time is taken up with loading and unloading.

Creep-feed grinding with continuous dressing has caused a revolution in the abrasives industry and is rapidly spilling over into other areas of machining and replacing the more pedestrian processes of milling, broaching, gear-cutting, and hobbing. High production facilities have capitalized on the very fast cut times associated with the process and have developed highly sophisticated, automated creep-feed grinding cell systems. Those cells have also been linked into fully automated factories, producing parts from raw castings/forgings/bar stock through to the finished part, inspected and marked with a part and

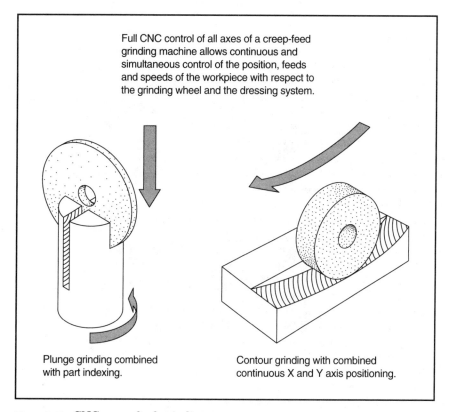

Figure 8.12 CNC creep-feed grinding.

serial number, with all recorded data stored for each individual workpiece. Creep-feed grinding is truly the process of the future.

8.4 Speed-feed Grinding

Speed-feed grinding is a most under-utilized process (see Fig. 8.13). It is a reciprocating grinding process, however, instead of the table running on at the end of each stroke, the table position end-stops are very carefully controlled, so that every table reversal takes place at precisely the same point. One of the main disadvantages of reciprocating form grinding is that the impact of the grinding wheel and the workpiece, for each stroke of the machine and for every part, severely chips away any precise form dressed onto the wheel periphery. Speed-feed grinding is an attempt to overcome that disadvantage. First, only one part is machined at any one time. The machine table is driven by a rigid toothed belt system, so that the position of the table reversals can be accurately controlled. The process is therefore set up so that the

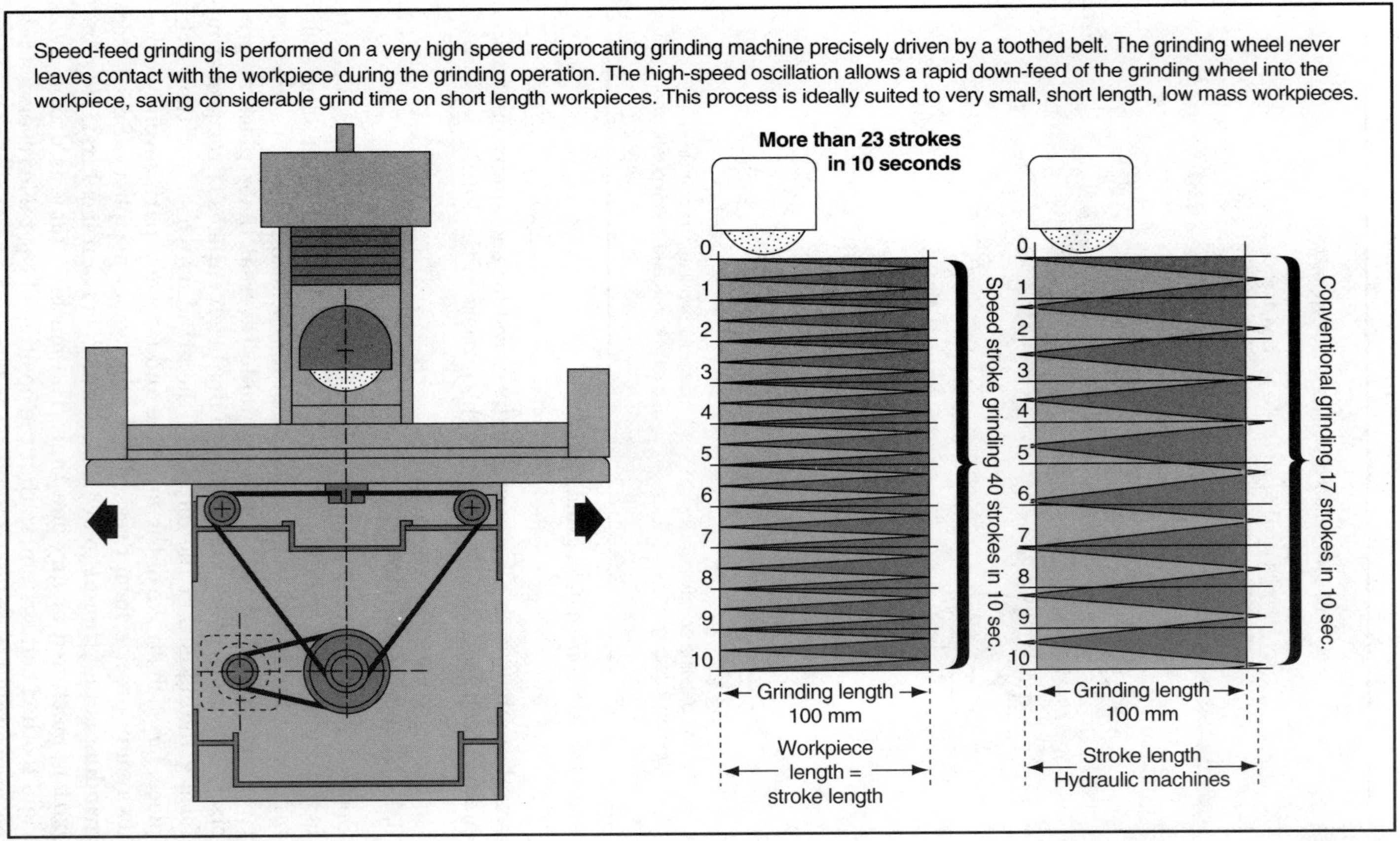

Figure 8.13 Speed-feed grinding.

grinding wheel never leaves the surface of the workpiece and suffers no impact forces. The table oscillates at a very fast rate, down-feeding at the end of each stroke just as the wheel is about to leave the surface of the workpiece.

This is a fast process and economizes on wheel usage. The types of parts suited to this process, however, are small, low mass parts, and those which are relatively short in table travel, e.g., cutting tool inserts, fuel control arms, and parts with a large depth/length of cut ratio.

Chapter 9

Special Grinding Processes

9.1 Centerless Grinding

Centerless grinding evolved from the needs of mass production. Centerless grinding allows a cylindrical workpiece to be brought into machining position very quickly and uses the inherent traversing motion of the process as the conveyance for the workpiece. Although the workpiece is not held or restrained by fixturing, centerless grinding is an exceptionally precise and consistent process.

The process uses two grinding wheels (see Fig. 9.1):

1. The regulating wheel, usually a rubber bonded grinding wheel which controls the speed of the workpiece during the grinding operation
2. The grinding wheel, which performs the cutting operation

In the through-feed configuration, the workpiece is placed between guides at the entrance between the grinding wheel and regulating wheel, supported on a work rest or blade. The grinding and regulating wheels rotate in directions that push the workpiece against the blade. The grinding wheel rotates in the range 30 to 40 ms^{-1} (6000 to 8000 sfm) and the regulating wheel in the range 25 mms^{-1} to 5 ms^{-1} (50 to 1000 sfm). The axis of the regulating wheel is set at an angle to the grinding wheel by about three degrees and can vary from almost zero to eight degrees. This angle determines the through-feed speed of the workpiece. The workpiece rest or blade is a support which has an angle of about 60 degrees to the horizontal and is oriented in such a manner to push the workpiece towards the regulating wheel. Although the centerlines of the grinding wheel and regulating wheel are in the same plane, the workpiece is usually set above that of the two wheels. This

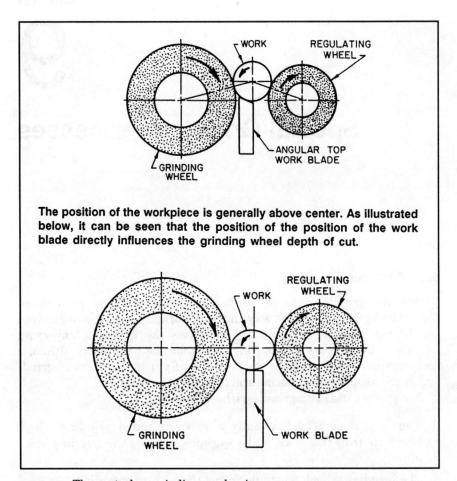

Figure 9.1 The centerless grinding mechanism.

distance rarely exceeds 12 mm (0.500 in). If the workpiece is too high between the wheels, chatter generally occurs as the workpiece tries to rise up and out from between the wheels.

In special cases where the workpiece is very long, the centerline of the workpiece can be below that of the grinding and regulating wheels to prevent whirling and whiplashing of the shaft.

There are other modes of centerless grinding, e.g., end-stop, which is a centerless operation where the regulating wheel holds the work against a stop so that a taper or a cylindrical form can be ground to a shoulder or a particular dimension along the shaft (see Fig. 9.2).

The work rest material is important as it can affect the surface finish of the workpiece. The material is chosen carefully: tungsten carbide for hardened steels and stainless steel; high-speed steel for nonferrous

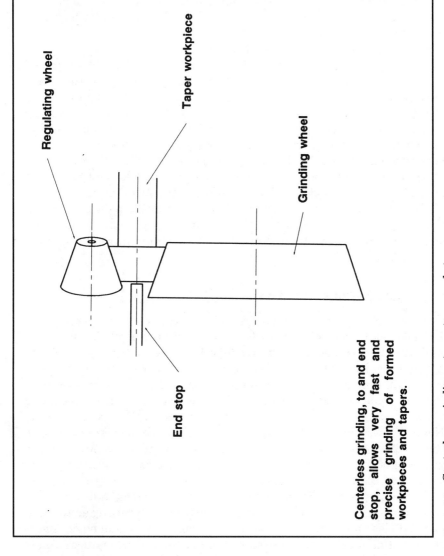

Figure 9.2 Centerless grinding a taper to an end stop.

parts; and cast iron for soft parts which might "pick up" and spoil their finish.

There is little concern for vibration or out-of-round in centerless grinding, as the workpiece is generally finding its own center and rounding with respect to the combined action of the two wheels and blade position. Centerless grinding is almost always used for high production grinding.

Recent studies into centerless grinding have shown that the asymptotic rounding-up relationship with time can be significantly improved with an adaptive control feature attached to the blade. As the workpiece is rotating between the two wheels, it is always trying to find its own center of rotation and, under the force of the regulating wheel, the largest amounts of out-of-round are pushed into the grinding wheel to be machined away. Remember, the regulating wheel is a rubber-bonded wheel, which, under the grinding forces, will deflect and only allow a portion of the out-of-round to be ground. The rest has to be ground off on the next rotation. Monitoring the grinding force on the blade and then positioning the blade either higher or lower between the wheels adaptively controls the rounding-up process. It has been shown that by this method, a significant reduction in rounding-up is achieved.

9.2 Tool and Cutter Grinding

Other machining processes require a cutting tool, e.g., turning requires a single point tool or form tool, milling requires a whole variety of milling cutters, broaching requires lengths of broach tools, and drilling requires drill bits. Each of these processes requires their tools to be ground, and in many cases reground or sharpened. In the case of turning and some milling applications, inserted carbide cutters are being used in order to cut faster and decrease resharpening costs. Nevertheless, the initial inserted cutters still have to be ground due to their extreme hardness and accurate form requirements. Much of the grinding of inserted cutters is performed by creep-feed grinding from the solid, diamond wheels. For milling cutters, the grinding and resharpening is a little more complex, as the shapes and forms are three-dimensional. A milling cutter not only has the cutting angles and back-off clearances, but also has a helix along the cutting edge. A special grinding machine was developed to machine these features and is called a tool and cutter grinder (see Fig. 9.3). The machine allows a grinding wheel, usually a cup wheel, to be fed into and along the length of the cutting edge as the cutter is rotated. The rotation is usually by hand and the whole process is most time-consuming. Worst of all, because of the complexity of the process in the movement and position-

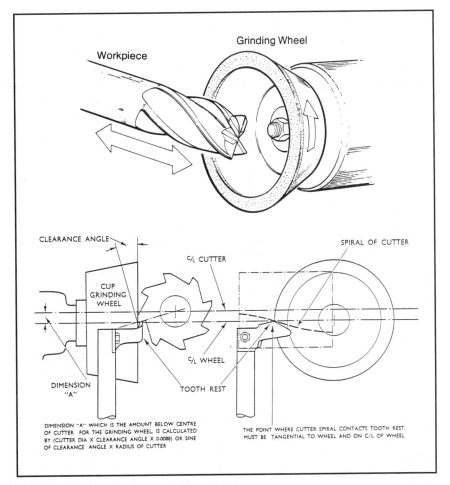

Figure 9.3 Tool and cutter grinding.

ing of the cutter, no cutting fluid is used. This allows the operator to see the grinding operation and so control the infeeds and clearance angles more accurately than if the operation were fogged out in a deluge of cutting fluid. Wear and misalignment of the finger rest occurs with the indexing of each tooth, so that the final geometry and tooth-to-tooth relationship is inferior and inaccurate. Such variation in the manual tool and cutter operation means that every cutter is effectively a different tool with slightly different geometry and wear characteristics. With the upsurge in CNC machining centers, consistency is the key where the manual method was unsatisfactory.

CBN grinding wheels opened up a new opportunity in this area because the grinding wheel does not appreciably change in size or form over a relatively long period of time. Because of the more consistent wheel geometry, the developing multi-axis CNC machine tool control systems could be tailored to accurately control the path of the grinding wheel in a fully enclosed machine. As there was now no manual intervention required, a copious flow of cutting fluid could be used to manufacture and resharpen cutters more consistently and with better quality cutting edges. There are no heat-checks or thermal damage to the cutting edges associated with burning from the lack of cutting fluid prevalent in the manual operation. The CNC tool and cutter grinder is being used increasingly across the industry. The use of superabrasives has indeed revolutionized the tool and cutter grinding business.

Such a machine requires many axes of movement in order to create all of the tool geometries. This makes the control systems not only very complex, but also quite costly. A CNC tool and cutter grinding machine may cost 30 times the price of a conventional, manual tool and cutter grinder. However, the versatility and the productivity, along with the consistency of the tools ground, very easily and quickly justify the added expense of the capital equipment.

9.3 Electrolytic Grinding

Electrolytic grinding is virtually not a grinding process, because, although a grinding wheel is used in the electrolytic grinding process, only 10 to 25 percent of the cutting action is abrasion (see Fig. 9.4).

Electrolytic grinding is the removal of material by the combination of electro-chemical decomposition and the abrasive action.

Typically a metal-bonded grinding wheel is used. The abrasive might be aluminum oxide or silicon carbide and on occasion CBN and diamond. The grinding wheel is mounted on a spindle which is insulated from the machine and connected to the negative side of a low voltage DC supply (4 to 16 V). The workpiece is connected to the positive side.

The abrasive media in the grinding wheel is not electrically conductive and protrudes beyond the metal bond surface of the grinding wheel periphery. These abrasive grains act as insulators between the workpiece surface and the grinding wheel proper. An electrolyte, a saline solution, is forced into the gap between the wheel and workpiece approximately 0.012 to 0.025 mm (0.0005 in to 0.001 in) and a high current is passed across the gap (50 to 3000 A). The combination of the current in the presence of the electrolyte causes an electro-chemical deplating action to take place. The electro-chemical action forms a soft oxide layer of decomposed material on the surface of the workpiece. The abrasive action simply removes the oxide layer to continuously expose a conductive surface. A compensating device maintains the gap distance so that a short does not occur and that the gap does not grow such that the current stops flowing.

Special Grinding Processes 191

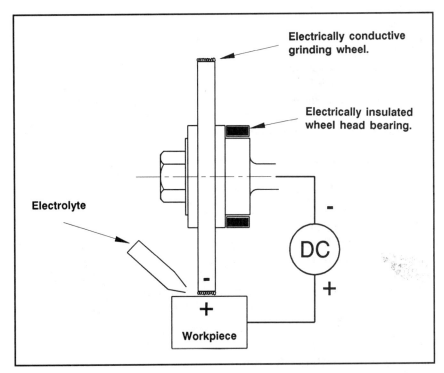

Figure 9.4 Electrolytic grinding system.

Prior to using a metal-bonded grinding wheel for electrolytic grinding, the current is reversed from the machining condition so that the grinding wheel is deplated to expose the abrasive grain. Three types of grinding wheels are used for electrolytic grinding.

Metal-bonded diamond wheels are usually used for flat grinding of carbides and very hard surfaces. These are very expensive wheels and are not suited to form grinding.

Diamond-plated wheels are used for form grinding. The form can be very accurately reproduced on the metal rim of the grinding wheel. Approximately 0.125 to 0.175 mm (0.005 to 0.007 in) of diamond grain is plated onto the wheel periphery.

The most widely used electro-chemical grinding wheel is the copper resin-bonded wheel. The grain is aluminum oxide and the wheel can very easily be formed with a single point diamond. It requires no deplating and has very good grain distribution throughout the grinding wheel.

ECG wheels are trued in the same manner as superabrasive grinding wheels; however, dressing is carried out by reversing the current and deplating the grinding wheel to expose the grain. The grinding wheels, from time to time, get clogged with the decomposed debris and have to be cleaned. This is accomplished by holding a pumice block against the wheel periphery to remove the loaded material, yet leave the grain undisturbed.

ECG has a stock removal rate proportional to the current used, typically 3 mm^3/s (0.010 in^3/min) for every 100 A used, depending on the type of material. Higher currents tend to cause worse surface finishes. Also, the arc length of cut is limited to no more than 20 mm (0.75 in), above which the electrolyte fails and leaves streaks and a poor surface finish.

ECG finds application in grinding very delicate parts, as the grinding process is virtually zero force. It leaves a part stress free and without burrs. It will grind very thin slots, leaving very thin wall sections without deformation or deflection.

Fixturing and handling of the workpiece is important as the electrolyte is a highly corrosive medium, though there are noncorrosive electrolytes on the market. The fixturing should be made from stainless steel. The workpieces should be thoroughly washed immediately after grinding to remove the corrosive electrolyte.

Surface finishes of 0.2 to 0.5 µm (8 to 20 microinches) can be achieved with ECG.

9.4 Honing and Super-finishing

Honing is the term given to the conditioning of a circular form, usually a cylindrical bore or shaft. Flat surface honing is termed "lapping" when the abrasive is a free abrasive. As the names imply, honing and super-finishing (see Fig. 9.5) are processes used primarily to impart a superior surface finish. However, in the automotive industry, cylinder bores are honed with quite a significant amount of stock being removed. In particular, when machining very long bores, honing is the fastest stock removal rate process.

Honing is carried out using a tool holder which holds a number of honing sticks. These sticks are abrasive blocks which have been manufactured so that they conform to the shape of the bore being honed. The process will not only impart a smooth surface finish, but also straighten bores which might previously be tapered, bell-mouthed, or barreled. It is important to carefully match the honing sticks with respect to their hardness grade. The "Grind-O-Sonic" device, described in Chap. 2, yields exceptional results and has led to significant improvements in the performance of the honing process.

The honing stones differ from grinding wheels in that the abrasive grain may be the same in sizes (80 to 600 grit) but they may also be manufactured with modified bonding agents, which include sulfur, resin, waxes, etc., to assist the cutting action.

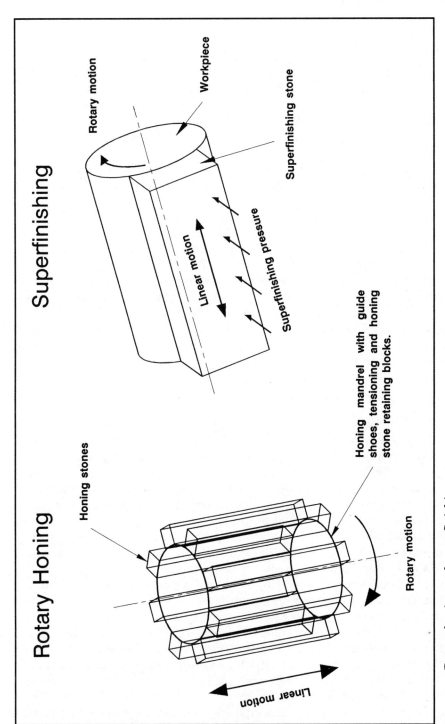

Figure 9.5 Rotary honing and superfinishing.

The honing stones are mounted in the holder. The holder is rotated as well as allowed to float to find its own center. Honing takes place at typically 250 to 1500 ms^{-1} (50 to 300 sfm). The hone is not only rotated, but also moved axially up and down the bore, so that the scratch patterns may be controlled. Very tight size control is achievable with honing, typically within 0.005 mm (0.0002 in).

Super-finishing is a variation of honing which conditions the outside diameter of a shaft. A honing or super-finishing block is pressed against the rotating shaft and removes a very small amount of material, smoothing out the peaks and valleys of the imperfect surface.

The action of super-finishing is based on the lubrication phenomenon that a lubricant of a given viscosity will establish and maintain a separating and lubricating film between two mating surfaces if their roughness does not exceed a certain value under a certain pressure. Under a given load, the super-finishing stone removes material until the smoothness of the surface creates a lubricating film and no cutting takes place. Therefore, under the control of the stone pressure, the surface of the shaft can be repeated based on the pressure used for the same given stone. Super-finishing is not only used to create smooth surfaces; it is also used to produce surfaces which have a well-defined scratch pattern, albeit very minute, to assist in lubrication.

9.5 Snagging and Cut-off

Abrasive processes, though they are able to remove large amounts of stock, appear to be somewhat "gentle" precision processes. Snagging, as the name sounds, is a very rough grinding process. It is usually used in heavy industry to dress castings from the mold or to descale steel ingots from the mill. These processes are usually hand operated and therefore take a great deal of punishment. The abrasive grain is very coarse and the bond systems forgiving.

Cut-off wheels is a term given to quite thin grinding wheels which are used as saws to cut runners and risers, for example, from castings. They can also be used, just like a metal-toothed saw, in general engineering operations like cutting metal bar stock, pipes, girders, etc. to length. Again, a property of the cut-off wheel is that the grain is tough and aggressive with a very forgiving bond system.

Cut-off wheels can cut at very high rates and leave an excellent surface on both the parent and the cut-off piece. There is a thermodynamic assistance for the cut-off operation that is present to a small extent in other grinding processes (see Fig. 9.6). As the cut-off wheel moves through the workpiece, there is a huge mass of material on either side of the wheel which conducts the heat away from the cutting zone. The material being cut acts as a coolant, dissipating the heat quickly and efficiently.

Special Grinding Processes 195

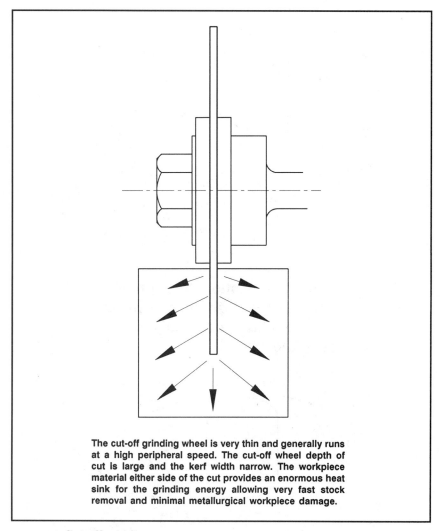

Figure 9.6 Cut-off grinding.

9.6 Double-disk Grinding

A more specialized grinding process is double disk grinding (see Fig. 9.7). Two grinding wheels are bolted to two opposing face plates and dressed flat. The wheels are brought together and the workpiece is fixtured into a rotary or piston fixture, which passes between the wheels. The workpiece is oscillated between the wheels as they are brought closer together to the finished size dimension of the part. Double disk grinding forms two flat and parallel surfaces on a relatively high production basis, depending on the fixturing, and load and unloading method of the workpiece.

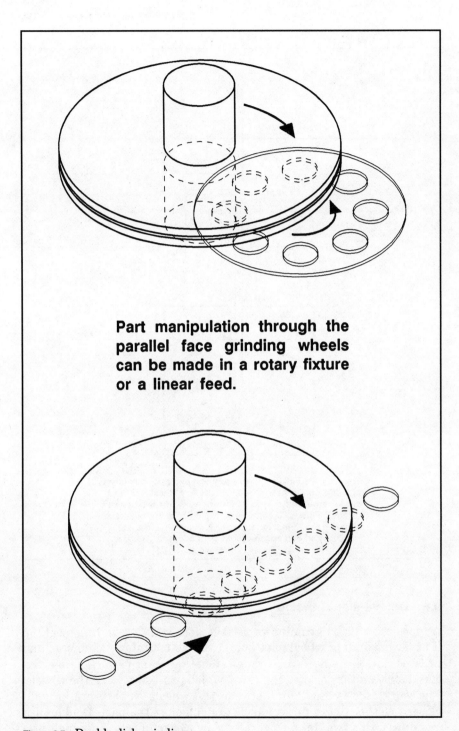

Figure 9.7 Double disk grinding.

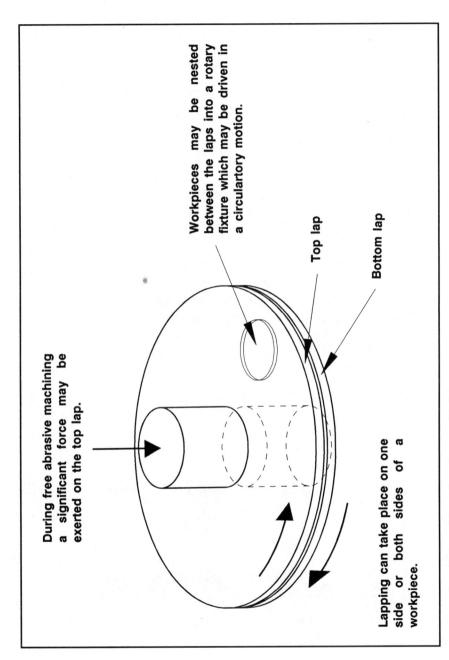

Figure 9.8 Lapping and free abrasive machining.

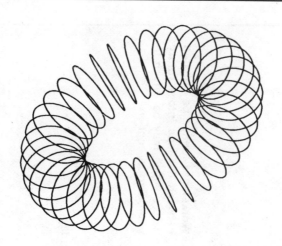

The toroidal shaped bowl is supported upon springs and rubber pads. An oscillating, motor driven, mechanism vibrates the bowl in a manner to excite the abrasive media to move in a spiral around the toroid. Workpieces are moved within the media under the vibratory motion and forces.

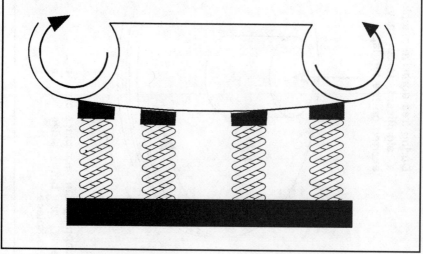

Figure 9.9 Vibratory finishing.

For a variety of applications the disk faces may be solid, slotted, or perforated for cutting fluid and swarf transportation.

9.7 Lapping and Free-abrasive Machining

There is very little difference between lapping and free-abrasive machining. Both processes produce flat surfaces to a high degree of surface finish. Lapping generally machines away a very small amount of material, merely imparting a surface finish to the workpiece, whereas free-abrasive machining can remove material at a high rate. Free-abrasive processes are termed free-abrasive since the abrasive grain is not bound by a rigid bonding or binding medium.

Free-abrasive machining refers to an abrasive suspended in a medium, usually oil and paste. In lapping (see Fig. 9.8), the lapping plates are usually made from relatively soft materials, the most popular being cast iron, though steel, copper, and aluminum can be used. These soft materials roll the very fine abrasive particles around and across the surface of the workpiece, but allow the abrasive grain to become imbedded into the surface of the lapping plate. The lapping plate effectively traps the abrasive, which truthfully is no longer a free-abrasive. Lapping achieves very high quality finishes and a high degree of flatness in the region of one light-band.

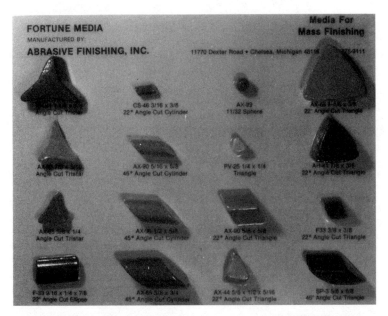

Figure 9.10 Examples of finishing media materials, size, and shape. (*Courtesy of Abrasive Finishing Inc.*)

Free-abrasive machining (see Fig. 9.8), in contrast, uses a very hard backup plate which allows all of the abrasive to roll between the plate and the workpiece. Under pressure from the top plate, the grinding force can be regulated to achieve stock removal. Heat can build up in the backup plate and cause undulations to occur, which effectively distort the plate and upset its precision, as well as degrade the cutting action.

One free-abrasive process is called "tumbling," which is the practice of literally shaking workpieces in a tub with an abrasive medium and deburring or surface conditioning the parts (see Fig. 9.9). There is little stock removed. However, the process which appears somewhat basic is in fact quite carefully controlled in the areas of the shape of the abrasive media, the tumbling action, and the forces generated between the workpiece and the abrasive. Tumbling and barrel finishing processes can process the roughest and most robust of parts to the most delicate and fragile workpieces, depending on the forces, media shape, size, and type (see Fig. 9.10).

Chapter

10

Coated Abrasive Processes

The manufacture, care, and safety of coated abrasive products is covered in Chap. 2. The market value of coated abrasive products is the largest of all abrasive sales.

Today, coated abrasives are competing readily with other grinding processes. It is not difficult to understand why. A coated abrasive belt is manufactured in a manner which orients the sharpest grain facets outermost. Belts can have incredibly long lengths, depending on the pulley arrangement. Thinking in terms of a plated grinding wheel, which also has only one layer of abrasive, the coated abrasive belt has a much larger "periphery," and therefore a correspondingly larger amount of abrasive to grind with. Belts generally run at a faster speed than the traditional vitrified grinding wheel, in the region of 50 ms^{-1} (10,000 sfm), as opposed to around 30 ms^{-1} (6,000 sfm) for vitrified wheels. A further benefit is the available width of an abrasive belt. Compared with a grinding wheel, which would have to make a series of passes and feed across the part, a belt can be exceptionally wide, allowing the machining of a very wide part in one pass, or a vast number of parts fixtured in a wide pattern. The only limitation on belt width is the maximum trimmed width of the production roll, 1,220 mm (48 in). It is not easy to make a belt larger than the maximum produced width.

The demand on the machine system then becomes a major factor, and as the need for higher horsepower increases, some machines exceed 185 kW (250 hp). Typically, most machines are less than 300 mm (12 in) wide. Belts 300 mm (12 in) wide or wider are considered wide belt machines and somewhat special.

The coated abrasive belt usually passes around two or more pulleys: a tensioning pulley and the contact wheel. The contact wheel is

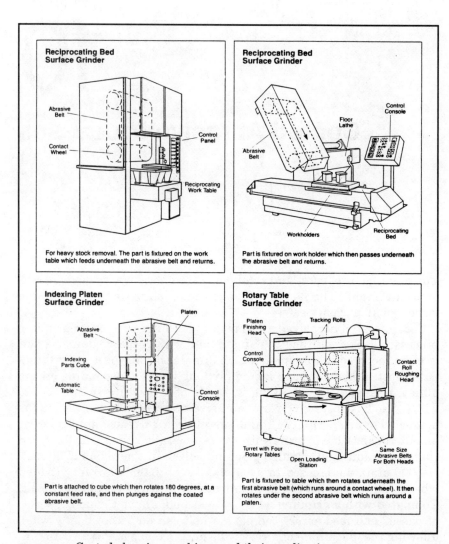

Figure 10.1 Coated abrasive machines and their applications.

Coated Abrasive Processes

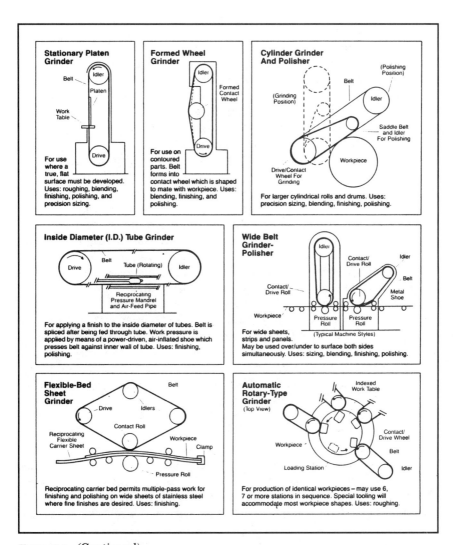

Figure 10.1 *(Continued)*

most important as it determines not only the cutting rate of the belt, but also the surface finish. The contact wheel is usually made of a rubber compound adhered to an aluminum wheel. The periphery of the wheel may be plain or serrated. The serrations are an important control over the aggressiveness of the cut. Large lands will make the abrasive act harder and more aggressive. Decreasing the land width makes the contact wheel act softer and conform more to the shape of the workpiece, which can also have the effect of a more aggressive cut. The angle of the serrations also affects the aggressiveness of the cut. As the serrations come closer to being perpendicular to the direction of rotation of the wheel, the cut is more aggressive. The contact wheel "tire" can be made from rubber, having a wide range of hardness measured by durometer. The combination of the hardness and the serrations in the contact wheel determine the cutting rate of the belt and the achievable surface finish. For a given belt, a hard wheel increases stock removal rate to the detriment of surface finish and a soft wheel gives a good finish for less stock removal rate. The contact wheel can be made in a whole range of hardnesses and densities. 90 A durometer is extremely hard and 20 A is extremely soft. The ranges are typically 70 to 90 A for stock removal, 40 to 60 A (medium range) for commercially acceptable finishes, and 20 to 30 A for very fine polishing with a limited contoured part. The pattern and size of the grooves in the contact wheel affect the process, influence the cutting action, and control the chip clearance to prevent belt loading. Some contact wheels are made of a highly compressed linen for better wheel workpiece conformance for polishing operations. The belts, used in combination with very soft contact wheels, have to be extra flexible and very tough. It is most important to store the contact wheels either flat or suspended in their bores. Storing a rubber wheel on its rim will cause a permanent set in the rubber and cause not only serious finish problems, but also the balance and effectiveness of the process.

There is a multitude of configurations of coated abrasive machines (see Fig. 10.1). The great majority of applications are for flat and plain cylindrical surfaces, generally with open tolerances. Coated abrasive processes are not typically super precise, due to the flexure of the contact wheel, the rigidity of the machine, and the give in the belt. Precision can be improved by using a steel contact wheel. Typical size tolerance ranges, at best within 0.05 mm (0.002 in) and typically within 0.125 mm (0.005 in). Highly complex forms are impossible to achieve with any reliability, however, some form grinding is possible using a formed platen or contact wheel (see Fig. 10.2). Such operations are usually combined with robot manipulation to ensure consistency.

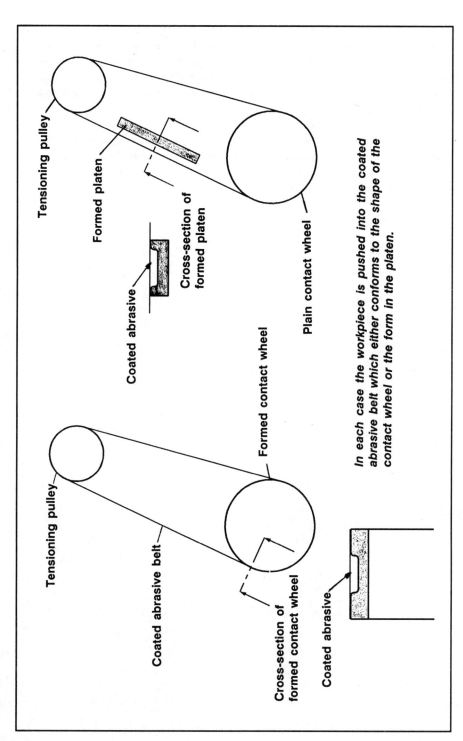

Figure 10.2 Form grinding with coated abrasives.

Coated abrasives are used in a multitude of hand-operated finishing machines to produce fine finishes on cabinetry in the woodworking industry and to prepare bodywork prior to paint spraying in the automotive industry. There are coated abrasive disks, flap wheels, and impregnated brushes, which all fall under the umbrella name of coated abrasives.

The 3M Company has recently developed an agglomerated aluminum oxide grain for use in coated abrasive products. In essence, the grain is a number of very small grains stuck together to make a larger grain. This gives the agglomerated grain a property similar to the microcrystalline CBN described on page 18.

Chapter 11

Surface Finish and Integrity Measurement

Abrasive machining is more traditionally regarded as a finishing process rather than a high-stock removal one. In this regard, abrasive machining processes produce the finest surface finishes of all machining processes.

Historically, surface finish represented how smooth a surface was. A good finish typically was one which was reflective, and a poor finish was one which was matte or dull. With respect to operational reliability, there is much more to a surface than whether or not it is "shiny." It is becoming more prevalent to categorize a surface with respect to its surface integrity. The important aspects of a surface are:

1. Profile—accuracy of dimension and form
2. Smoothness—surface finish, including the lay of the machining marks, the distribution of peak to valley height, and the pattern of superficial surface textures
3. Residual Stress—susceptibility of the surface to crack under mechanical and/or thermal cycling
4. Metallurgical and/or Sub-surface Damage—grain growth, changes in hardness or surface metallurgical structure

Grinding is a process where many cutting edges perform individual tasks of removing material. The characteristics of those cutting edges can be modified or changed by conditioning the abrasive with the proper truing and dressing methods. If the abrasive grain becomes dull, it will burnish the surface of the workpiece and generate a great deal of frictional heat. Some of that heat will be conducted into the surroundings; however, the great majority will be conducted into the workpiece surface. The heat may be detrimental to the process, in that the increase in surface temperature of the material may cause it to expand. Sudden cooling, or quenching, of the workpiece surface by the cutting fluid might cause the surface to contract at a very fast rate. This rapid contraction will usually generate high tensile residual stresses and may even cause cracks to appear in the workpiece surface. Even if cracks were not evident, a very brittle Martensitic structure might occur in the surface layer of the material. The heat from grinding might also cause the surface to simply heat up. The increase in surface temperature could induce an annealing of the surface and create a soft spot or grain growth.

Other than an obviously severe crack, these detrimental surface conditions can pass undetected by the naked eye. The grinding of ceramics and glasses, for example, can result in the desired surface finish and be crack-free. However, a poorly controlled abrasive process may cause sufficient sub-surface damage resulting in part failure while in service.

Scanning Electron-beam Microscopy (SEM) is a science which can perform an exacting analysis of a surface and sub-surface in order to determine its integrity. SEM analysis uses the bombardment of an electron stream onto a surface in a vacuum to produce a very accurate and detailed stereo representation of the workpiece surface (see Fig. 11.1).

Each chemical element has a definitive ionization energy. That is the energy, measured in electro-volts, required to remove the outermost electron from a positive charged atom of that element, called an ion. The SEM can measure the electron-volt potential, scan the surface of a material, and plot its material make-up. If, during grinding, certain undesirable elements of an alloy migrate to the grain boundaries and weaken or damage the metallurgical structure, it can be spotted at once.

Casual inspection of a surface will not always reveal cracks, as they may be extremely small or may have been burnished over by the rubbing action of the dull abrasive grain. Etching, a metallurgical technique which uses acid solutions to remove a very minute amount of the surface material, is used in order to expose surface cracks. In cases of critical surfaces, like aircraft engine parts, crack detection can be taken a step further by treating the etched surface with a dye penetrant and inspecting it under ultraviolet light. The dye seeps into the cracks. After a dusting with a fine powder, the dye then leaches out and becomes highly visible under ultraviolet light.

Etching and binocular inspection can reveal surface grain growth and structural changes created by the surface heat treatment, which may have occurred from the heat generated in a poorly controlled

Figure 11.1 Scanning electron micrographs of CBN grain.

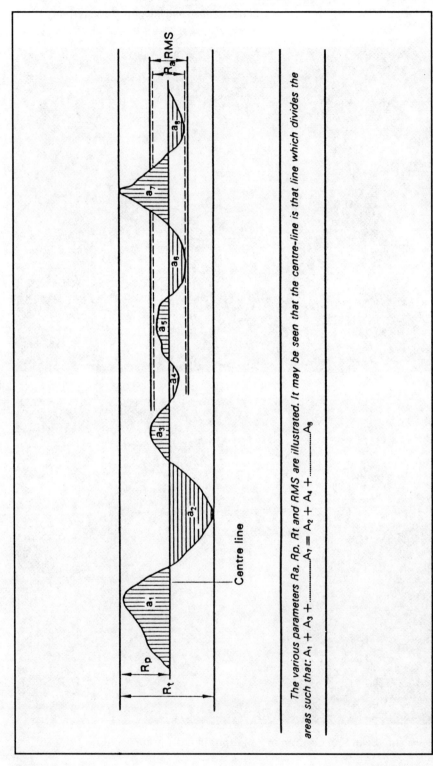

Figure 11.2 Surface finish measurement.

Surface measurement parameters

Nominal description	Profile description	Symbol	Definition	Unit
Waviness and roughness	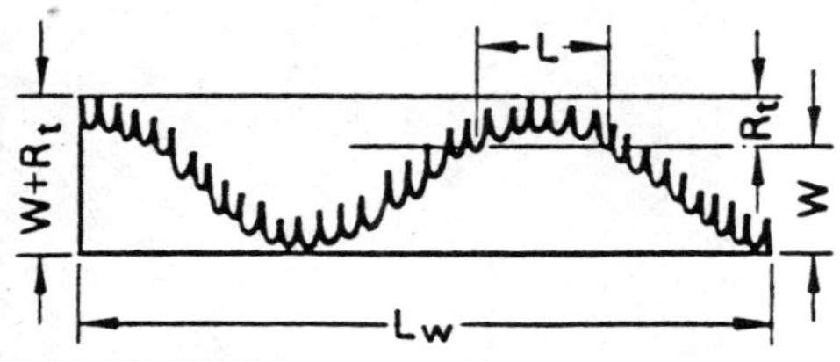	$W+R_t$	Total profile depth measured over L_w	μm
Waviness		W	$W+R_t$ Less average roughness depth	μm
Roughness		R_t	Total roughness depth over L	μm
Smoothing depth	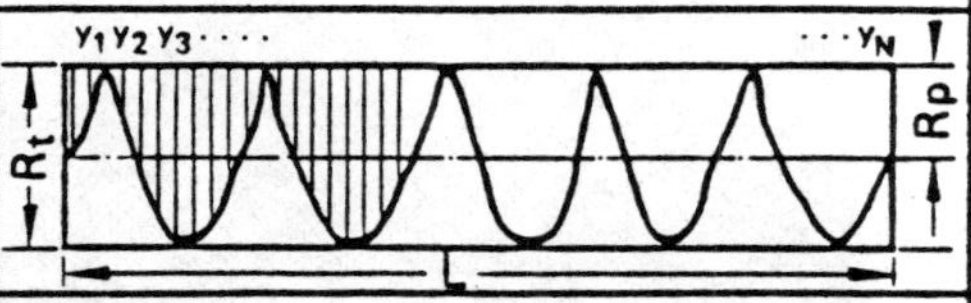	R_p	$\dfrac{1}{L}\int_0^L |y_i|\,dx \approx \dfrac{y_1+y_2+\cdots y_N}{N}$	μm
Arithmetical Average		R_a = CLA = AA	$\dfrac{1}{L}\int_0^L |h_i|\,dx \approx \dfrac{|h_1|+|h_2|+\cdots|h_N|}{N}$ $1\,\mu m = 40\text{ Microinch}$	μm μin

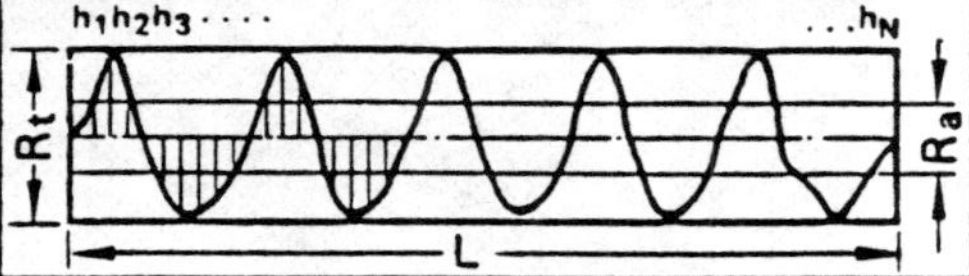

Figure 11.2 *(Continued)*

Both of the above surface pictures are at 100 magnification and clearly show the smearing and pitting of the surface as well as the random nature of the topography all of which goes unnoticed by the "one scratch" surface measurement technique.

Figure 11.3 Working surfaces.

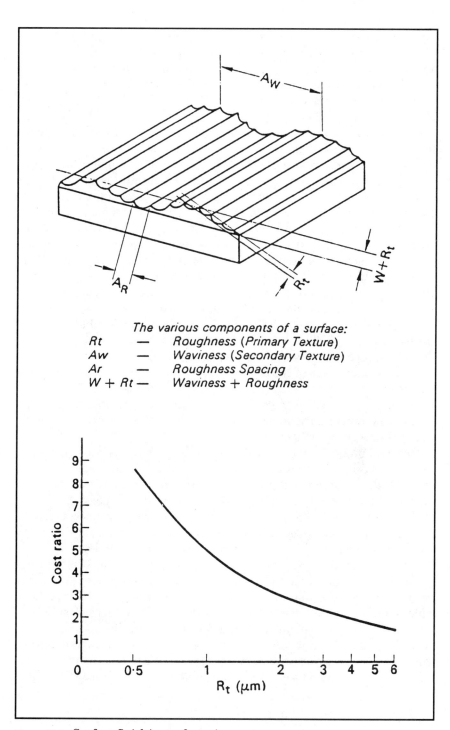

Figure 11.4 Surface finish/manufacturing cost.

grinding operation. There are other crack detection techniques available like X-ray, Neutron Radiography, and magnetic crack testing. Visual binocular and dye penetrant testing, which are both nondestructive testing methods, along with the combination of SEM analysis, which is a destructive test (as only small samples of material can be analyzed in an SEM), are the most accurate.

Cracks and metallurgical imperfections are somewhat obvious faults. However, studies show that the pattern of the machining marks or redeposited material can seriously affect the safe operation of certain materials and components. Abrasive machining is generally the last machining process prior to a workpiece entering service. It is therefore critical that the surfaces produced are of the highest quality.

Surface finish measurement, up until recently, has been made by tracing a stylus at 90 degrees to the machining marks across the ground surface. A calculation is then made by the surface finish measuring machine, which is the value for the mean height deviation from the nominal surface, termed R_a, or arithmetic average (see Fig. 11.2). Other numbers can be displayed: roughness (R_t), smoothing depth (R_p), average peak to valley height (R_{tm}), machine measurable, and (R_z) measurable graphically only. Such measurements are good guides, but do not represent the overall surface texture. The radius of the stylus and the physical size of the stylus do not allow it to follow, exactly, the surface profile characteristics. At best, the trace is good for only the short linear "scratch" it makes over the surface. This type of measurement is no longer satisfactory; in fact, it tells the engineer very little about the working surface (see Fig. 11.3).

New techniques are now available to assess an entire area of the surface of a workpiece. Replication of the surface can be made and carefully studied, giving a more realistic measurement of the working surface. The replication method takes a piece of polymeric substance, which temporarily adheres and conforms exactly to a surface. Once the polymer has cured, it is peeled from the surface and analyzed in an SEM, or plotted graphically using laser mapping technology.

The increasing importance of the working surface in the field has given rise to much research and development in the area of surface technology. Future manufacturing process engineers will have to be able to define and assess the surfaces they produce more closely than at present and, therefore, be more familiar with the inspection and measurement techniques. The burden is also on the design engineer, who specifies a surface finish/integrity value. The engineer's decision will affect the product's economical production as well (see Fig. 11.4).

Glossary

Abrasive An abrasive is a material which is very hard. It may be naturally occurring, e.g., diamond and quartz, or synthetic, e.g., aluminum oxide, silicon carbide, and CBN. An abrasive has to be harder than the material it is machining.

Aluminum Oxide Aluminum oxide (Al_2O_3) is an abrasive refined from naturally occurring bauxite. Aluminum oxide is used in a form of grain crushed from the fused alumina. Aluminum oxide may also be used in a pure crystal form.

Angle Approach Grinding Angle approach grinding is a cylindrical grinding method whereby the grinding wheel is plunged into the workpiece at an angle. A dressed form on the grinding wheel may grind both a number of side faces and a number of diameters in one plunge.

Arc of Contact The arc of contact is referred to as the distance from the top to the bottom of the arc which contacts both the workpiece and the grinding wheel.

Area of Contact The area of contact is referred to as the arc of contact multiplied by the width of the grinding wheel. This is not, however, the true area of the grinding wheel in contact with the workpiece. The area of abrasive grain in contact with the workpiece is typically 0.5 to 5 percent of that area.

Balance (Dynamic) Dynamic balance is the balance of the complete rotating grinding wheel and arbor mounted on the machine spindle measured at wheel speed.

Balance (Static) Static balance is the balance of the grinding wheel mounted on an arbor, typically balanced on a pair of knife edges.

Ball-screw A ball-screw is a very precise screw where the profile of the helix is a bearing race. The nut which runs on the screw has highly loaded ball bearings between the nut body and the race. Such a device is used for the precise, low-drag transmission of motion of machine slideways. Ball-screws are best used at slow speeds.

Base The base of a machine is the massive structure which provides the support and stability for the machine guideways. Traditionally, the material for the base has been cast iron; some are steel weldments. Lately, machines have been constructed of epoxy concrete.

Bauxite Bauxite is a naturally occurring mineral rich in alumina used in the making of aluminum oxide.

Blotters Blotters are thin paper or soft plastic washers which must be placed between the sides of the grinding wheel and the flanges of the arbor to prevent the wheel from cracking.

Bond Bond material is the material which holds the abrasive grain in place in the form of a grinding wheel.

Burning Workpiece burning is the blackened discoloration which indicates very poor grinding conditions. The discoloration is caused by surface oxidation due to extremely high surface temperatures.

Burnishing Burnishing is the very heavy rubbing of two surfaces, where one is much harder than the other. The result is that the softer surface is flattened and has a very shiny appearance. Typically the surface is highly stressed.

Burr A burr is the ragged material which is turned over the edge of a cut. Burrs are particularly prevalent with blunt cutting tools and those generating high surface temperatures.

Bushing A bushing is the insert sometimes found on the bore of a grinding wheel. The insert bushing may be lead, wax, or plastic.

CBN Cubic Boron Nitride; a superabrasive.

CNC Computer Numerical Control. A computer is used to control the movement of the axes of the machine tool.

Cam Grinding Cam grinding is the machining of the cam profile by grinding. In this process the cam is moved back and forth with respect to the grinding wheel, while the cam is rotating to create the cam profile.

Center A center is the conical support for a cylindrical workpiece. The center is a cone of a 60-degree angle which may be solid (dead center) or rotate on a bearing (live center).

Centerless Grinding A cylindrical grinding process using two wheels, one which removes the material and the other which controls the speed of the workpiece.

Centrifuge A centrifuge, or cyclone, is a cutting fluid filtering system which separates solid particulate from the fluid by spinning it to allow centrifugal force to separate the heavier particles. The centrifuge motor is driven at high speeds, whereas the cyclone is driven by the path of the fluid in a spiral chamber.

Chatter Chatter is the audible vibration caused by a poor grinding process. The grinding wheel may be dull or loaded with material. The machine may be weak. The workpiece may be fragile and flimsy. Regenerative chatter is a special form of chatter.

Chip The chip is the material removed by one cutting edge or grain.

Cleavage Plane A plane of crystallographic weakness in a material along which preferential shear takes place.

Coated Abrasive Abrasive product which is adhered to cloth or paper.

Concentration   The amount of superabrasive material contained in a unit volume of the grinding wheel. The measurement is based on the number of carets per unit volume, e.g., 100 concentration is 4.4 carats/cm^3 or 72 carats/cu. in.

Contact Wheel   Usually an aluminum wheel with a solid hard rubber tire used as the back-up to a coated abrasive belt.

Continuous Dressing   Continuous dressing is the process where a diamond dressing roll dresses the grinding wheel all the time the grinding wheel is machining the workpiece.

Coolant   Coolant is the misnomer for cutting fluid. The cutting fluid does not only cool; it lubricates, cools, and washes away chips and debris. The better term is cutting fluid.

Corundum   Corundum was one of the early naturally occurring abrasives.

Creep-feed Grinding   Creep-feed grinding is a process where a very soft and porous grinding wheel is used to take a deep cut in one slow pass. The process has high stock removal and very accurate retention of form. It is particularly used on difficult-to-machine materials.

Cross-slide   The cross-slide on a surface grinding machine is the axis at right angles to the table and parallel with the machine spindle.

Crush Dressing/Crush Forming   Crushing is a technique where a steel or tungsten carbide roller with a form is pushed into a slowly rotating grinding wheel to form the profile shape onto the grinding wheel periphery.

Cubic Boron Nitride   CBN; a superabrasive mineral second only in hardness to diamond.

Cut-off Wheel   A cut-off wheel is a thin grinding wheel, usually resin- or rubber-bonded, used just like a circular saw.

Cutting Fluid   Cutting fluid is the fluid used to cool, lubricate, and wash clean a process. Cutting fluids may be oils, water-based synthetic, or soluble oils and gases.

Cylindrical Grinding   Cylindrical grinding is a process for the machining of round components.

Deburring   Deburring is the removal of burrs from the corners of workpieces.

Detritus   Grinding detritus is the residue from the metal removal process, typically the chips and the free/loose abrasive.

Diamesh   A patented process whereby superabrasive grains can be placed in an exact regular pattern on the surface of an abrasive tool.

Diamond   Diamond is both a naturally occurring and synthetic substance. It is the hardest substance known to humans.

Diamond (Single-Point)   A single-point diamond is a single diamond mounted into a steel holder. The diamond is like a stylus and is used for dressing the periphery of a grinding wheel.

Diamond (Multi-Point)   A multi-point is a steel post which has small diamonds impregnated on the end of the post by sintering or plating. The multipoint is used for dressing the periphery of grinding wheels.

Diamond Roller Dresser   A diamond roller dresser is a wheel which has a surface which has been impregnated with diamonds either by plating or sintering. The surface of the roller may be formed into a shape for profiling a grinding wheel. Diamond roller dressers achieve the most accurate detail of form profiles and have a very long life.

Dog Drive   A dog drive (sometimes called a pin and carrier drive) is a mechanism which engages a plain cylindrical workpiece and is supported only by two centers.

Down-cutting   Down-cutting is the term in creep-feed grinding when the grinding wheel is rotating in a direction which, at the arc of cut, is the same as the motion of the workpiece. This is sometimes termed climb grinding.

Dresser   A dresser is a device used to condition the periphery of a grinding wheel.

Durometer   This is a measurement of the hardness of rubber. The higher the number, the harder the rubber.

Emery   Emery is a naturally occurring abrasive.

Encoder   An encoder is an electronic device which is used to accurately divide one revolution into a finite number of parts for motion control.

External Grinding   External grinding is the grinding of the outside diameter of cylindrical components.

Feedback   Feedback is the signal sent back to a control device, which is used to provide verification and/or corrective action.

Feed lines   Feed lines are the unwanted effect from feeding a grinding wheel down a cylindrical roll, leaving a "barber pole" pattern in the surface of the roll.

Floor-to-floor Time   Floor-to-floor time is the time taken to perform an operation, from the time the part is picked up from the floor to when it is completed and put down on the floor. This time includes the cutting time, the loading time, and unloading time.

Flour Wheels   Flour wheels are extremely fine-grain wheels in the region of 4000 grain size.

Free-abrasive Machining   This is a process where a loose abrasive is suspended in a fluid and used between two hard, flat, rotating surfaces to machine a flat surface. It is a process similar to lapping, but removes material at a faster rate.

Friability   Friability is the ability of the abrasive to fracture.

G-Ratio   G-Ratio is the ratio of the amount of workpiece material removed versus the amount of grinding wheel used.

Glazing   Glazing is the term given to a grinding wheel periphery which has become worn flat by a very hard workpiece.

Grade The grade of a grinding wheel usually refers to its hardness, the abrasive type, the grain size, the bond material, and its porosity. There is a standard code for grinding wheel grades.

Grain Size The grain size refers to the number which corresponds to the mesh size used for sizing the abrasive grain.

Green State A grinding wheel is said to be in a green state before it is fired. Its only means of being held together is the water in the mix.

Grit Another term for abrasive grain.

Guideways The guideways of a machine are the very precise linear bearings which guide the axes of movement.

Headstock The headstock is the part of a grinding machine which houses the grinding spindle.

Heat Flux Heat flux is the term given to the amount of heat generated over the area of contact between the grinding wheel and the workpiece.

Hydrodynamic Hydrodynamic is referred to in connection with high-speed slideways where the slideway may rise up on the lubricating oil film when in motion. An example is a water skier who rides on the water in motion but sinks when stationary. The hydrodynamic principle is also used in very precise bearing applications. In these systems the bearing is only supported or levitated when it is in motion.

Hydrostatic Bearing A hydrostatic bearing is an extremely stiff bearing which works on the principle of pockets of very high pressure oil. The pockets oppose one another, such that if the inner race begins to deflect, then the pocket area on one side is constricted and on the other it is opened. The oil pressure then acts to keep the inside and outside concentric.

ID Grinding ID grinding is the grinding of the inside diameter of holes or profiles using a very small, yet very high RPM grinding wheel.

Infeed The infeed or down-feed is the feed motion of the grinding wheel into the workpiece.

In-process Gage In-process gaging is a mechanism which measures the size of a component during or directly after the machining cycle, so that very accurate sizing may be achieved without removing the workpiece from the machine.

Lapping A process which uses a fine loose abrasive in a fluid suspension to produce very fine finishes and a high degree of flatness.

Loading Loading is a term given to material which has become attached to the periphery of the grinding wheel.

Martensitic Structure A martensitic structure is a very hard and brittle structure which increases in hardness with increasing carbon content in steels. This structure occurs after heating and rapid quenching.

Maximum Normal Infeed Rate The feed rate of the workpiece normal to the grinding wheel at the top of the arc of cut.

Mesh Number The mesh number refers to the fraction of an inch that the wire mesh is spaced for sizing the abrasive grain.

Micro-inch One millionth of an inch (0.000025 mm or 0.000001 in).

Micron One millionth of a meter (0.001 mm or 0.00004 in).

Norbide Norbide is a trade name owned by the Norton Company for a very hard material used in stick form to roughly hand dress grinding wheels.

OD Grinding OD grinding is the cylindrical grinding of an Outside Diameter. See external grinding.

Pantograph A pantograph is a system of levers which allows the magnification or the miniaturization of a profile to be drawn.

Paradichlorobenzene Paradichlorobenzene (moth ball crystal) is a compound used to induce high porosity into vitrified grinding wheels.

Particulate Particulate is a very small particle suspended in a fluid.

Plunge Grinding Plunge grinding is a process where the grinding wheel is driven radially into a cylindrical workpiece to finished size with no cross-slide movement.

Porosity Porosity is the void within a grinding wheel which makes it appear sponge-like.

Regenerative Chatter See section on cylindrical grinding.

Resinoid A resinoid bond is a bond which is a thermosetting synthetic polymer, usually designated B.

Resolver A resolver is a device like a stepper motor, which may be given a signal from an electronic control to rotate a ball-screw a set number of turns in order to move a slideway. A resolver has no feedback.

Resonant Frequency Resonant frequency is the frequency of vibration of a system which creates the largest amplitude.

Rest A steady rest is a work support device used to support long, slender workpieces in cylindrical grinding.

Rigidity Rigidity is the stiffness of a system and is measured by the deflection of the system under a given load in units of Nm^{-1} (lbf/in).

Ringing A means of checking a vitrified grinding wheel for cracks prior to mounting it on a grinding spindle. The wheel is struck firmly, but gently. A clear ring signifies a good wheel and a dull ring a suspect wheel.

Saddle The saddle of a surface grinding machine is the part which houses the cross-slide and the table ways.

SEM Scanning Electron Microscope.

Silicon Carbide (SiC) A synthetic abrasive, usually green or black in color.

Spark-out This is the grinding of a workpiece at the end of a grind cycle without engaging any further downfeed. The grinding forces are allowed to subside with time, ensuring a precision surface.

Specific Energy   Specific energy is the amount of energy used to remove a unit volume of material in units Jmm^{-3} (Btu/in^3).

Steady (Steady-rest)   A steady is a supporting device for cylindrical workpieces to prevent their deflection under the force of themselves or the grinding action.

Stiffness   Stiffness is the resistance of a system to deflect under load. Static stiffness is the stiffness measured under stationary conditions. Dynamic stiffness is a stiffness measurement made at working speeds.

Stock   Stock is an amount of workpiece material to be removed.

Stock Removal Rate   Stock removal rate is the rate at which material is removed from a workpiece.

Structure   The structure of a grinding wheel is the relative size, proportion, and position of the abrasive grain, bond material, and porosity in a grinding wheel matrix.

Superalloy   A superalloy is a high nickel- or cobalt-based alloy, typically used in aerospace applications and high-temperature engines.

Superfinishing   This is a process which can produce the highest surface finishes in round components by exerting considerable pressure on a shaped honing stone to impress it into the rotating cylindrical surface.

Surface Finish   Surface finish is the smoothness of the machining marks on the surface of a workpiece. Surface finish is measured by scratching a precision stylus across the surface and measuring the amplitude of the fluctuations of the stylus.

Surface Grinding   Surface grinding is the machining of a flat, angled, or contoured surface by passing a workpiece beneath a grinding wheel in a plane parallel to the grinding wheel spindle.

Swarf   Swarf is the mass of chips and debris remaining after a grinding process.

Table   The table of a surface grinder is the part of the machine which supports the workpiece and reciprocates back and forth beneath the grinding wheel.

Tailstock   The tailstock is a part of a cylindrical grinding machine which provides center support to a workpiece. The tailstock is situated on the workslide opposite the headstock/workhead.

Thermal Damage   Thermal damage is metallurgical damage which occurs to a material when it is subjected to temperatures which will affect its metallurgical structure. Thermal damage often occurs before the visible burn marks.

Thermal Stability   Thermal stability is the stability of a system to remain within tolerance during a fluctuation in temperature.

TIR   Total Indicator Reading. A tolerance of +0.01, –0.01 has a TIR of 0.02.

Tolerance   An exact dimension can never be achieved. Therefore, an amount of error either above and below, all above, or all below the desired size is given. This is the tolerance.

Truing Truing a grinding wheel is the method of making the grinding wheel rotate concentrically to the center of rotation of the spindle.

Truing Brake A truing brake is a grinding wheel mounted in an arbor in a bearing house, which has a friction brake to limit its speed. Such a truing brake is used to true resin-bonded superabrasive grinding wheels.

Up-cutting Up-cutting is the term in creep-feed grinding when the grinding wheel is rotating in a direction which, at the arc of cut, opposes the motion of the workpiece.

Velocity Profile Velocity profile is the velocity that a cam follower has during the revolution of a cam.

Vibrational Stability Vibrational stability is the stability of a system to remain within tolerance across a range of vibrational frequencies.

Vitrified A vitrified bonding system is one which is very brittle and is achieved by firing a mixture of abrasive and feldspar or silica in a kiln, like a piece of pottery.

Wheelhead The wheelhead is the part of a grinding machine which houses the grinding spindle.

Wheel Speed Wheel speed is the peripheral speed of the grinding wheel.

Workhead The workhead or headstock is the part of a cylindrical grinding machine which holds and rotates the workpiece.

Work Speed Work speed is the rotational or translational speed of the workpiece while being machined.

Index

Abrasives
   Aluminum oxide, 12
   Ceramic, 14
   Coated, 3, 34, 201
   Cubic Boron Nitride (CBN), 15
   Diamond, 19
   History, 3
   Market, 6
   Silicon carbide, 9
Absolute programming, 119
Adaptive Control (AC), 101
Adhesives, 36
Agglomerated grain, 205
Angle approach grinding, 139, 149
Arc length of cut, 90
Area of contact, 97
Axis drives, 161

Backing materials, 36
Balancing, 42
   Manual 3 weights, 46
   Manual 2 weights, 47
   Automatic with fluid, 48
   Automatic with weights, 49
Belts
   Coated abrasives, 34
Blanchard grinding, 172
Blotters, 41
Boiling, nucleate and film, 123
Bonding
   Plated, 31
   Resin, 30
   Rubber, 31
   Metal, 31
   Vitrified, 26
Brake true dresser, 52

Cam shaft grinding, 153
CBN (Cubic Boron Nitride), 15

Centerless grinding
   Thru-feed, 185
   Taper, end-stop, 187
Ceramic
   Abrasives, 14
   Machining, 100, 103
   Market, 7
Chatter
   Regenerative, 152
   Vibration, 104, 110, 171
Coated abrasives, 34, 201
   Fitting a belt, 50
   Robotic part manipulation, 204
Computer Numerical Control (CNC), 116
Concentration, 23
Continuous dressing, 85
   Dresser infeed rates, 86
Contour grinding, 181
Coolant, 121, 129
Creep-feed grinding, 175
   Machines, 106
   Process, 175
Cubitron, 14
Cut-off, 194
Cutting fluid, 121
   Application, 126
   Damming, 128
   Filtration, 127
   Refrigeration, 129
Cylindrical grinding, 133

Diamesh, 32, 92
Diamond roll dressers, 71
   Critical dwell, 81
   Mounting, 74, 82
   Speed ratio, 79
   Wear, 83
Dog drive, 137

Index

Double disk grinding, 195
Dressing, 50
  CNC, 55
  Crush, 64
  Diamond roll, 67
  EDM, 32, 85, 87
  Fliese, 61
  Forces, 66
  Overlays, 63
  Roll-2 Dress, 56
  Single point, 55, 60
  Swing-step, 78
  Wafer roll, 65
Durometer, 204

Electro-Discharge Machining (EDM) dressing, 32, 85, 87
Electrolytic grinding, 190
Energy in grinding, 94
  Cutting, plowing and rubbing, 95
Epoxy granite machine bases
  Granitan, 110

Feed-lines or "barber pole" effect, 142
FEPA, 9, 10, 25
Film boiling, 123
Fliese tools, 61
Form grinding, 172
Free-abrasive machining, 199
Friability, 13, 22

G-Ratio, 16
Grading system
  Wheels, 10
  Belts, 38
Grain size, 24
Granitan, 110, 165, 170
Grind-O-Sonic, 26, 192

Hardness
  Abrasives, 21
  Wheels, 26
  Durometer, 204
High-speed grinding
  Machines, 109
  Process, 126
Honing
  Centers, 138
  Process, 192
Hydro-static bearings, 168

ID (Inside Diameter) grinding, 153
Incremental programming, 119
In-process gaging, 157

Lapping, 199

Magnetic bearings, 109
Maximum Operating Speed (MOS), 28
Micro-milling analogy, 89

Nano-precision, 5
Next Generation Grinding Machine (NGGM), 112
Nickel coating, 17
Normal force, 99
Nucleate boiling, 123

OD (Outside Diameter) grinding, 133
  Feeds and speeds, 142
Overlays, 63

Plating
  Wheels, 31
  Dressers, 71
Plunge grinding, 181
Power monitoring, 99
Pre-loaded roller guideways, 164

Reciprocating grinding, 161, 177
Residual stress, 207
Resin bond, 30
Rigidity, 107, 113
Ringing, 41
Roll grinding, 135
Rotary grinding, 172
Rotary table machines, 115
Rubber
  Bond, 31
  Machining, 11

Scanning Electron Microscopy (SEM), 208
Seals
  Reaction with cutting fluid, 132
  Surfaces, 149

Sensors, 99
SG (Seeded-Gel or Sol-Gel), 14
Silicon carbide, 9
Snagging, 194
Specific energy, 94
Speed-feed grinding, 181
Sticking, 54
Stiffness, 107, 113
Storage
    Wheels, 34
    Belts, 39
Superabrasives
    Diamond, 19
    CBN, 15
Superfinishing, 192
Surface finish measurement, 207

Thermal stability, 110, 129
Tool and cutter grinding, 187
Toughness, 22
Truing, 50

Tumbling (barrel finishing), 199

V-flat guideways, 166
Vector feed rate, 120
Vertical spindle grinding, 172
Vibrational stability, 105, 110, 171
Vitrified bond, 26
    Structure, 29

Wafer roll dressers, 65
Wear of abrasives, 20, 95
Weldment, 110, 165
Wheel sharpness
    Surface finish, 122
    Heat, 99
    Surface integrity, 207
    Forces, 99
Wheel tightening/mounting, 43
Wheel wear sensing, 99
Work rests or steadies, 143

ABOUT THE AUTHOR

Stuart C. Salmon, Ph.D., is president of Advanced Manufacturing Science and Technology, Inc., an independent consulting firm in Cincinnati, Ohio, specializing in the areas of grinding, abrasives, advanced machine tool design, and manufacturing technology. He is known worldwide as the chief contributor to the research and development of the revolutionary Continuous-Dress Creep-Feed grinding technique. Dr. Salmon took his apprenticeship from and was sponsored for his grinding research by Rolls Royce Aeroengine Division in England. He has also worked for the General Electric Aircraft Engine Group in Evandale, Ohio, and has written numerous technical papers and articles, including contributions to the *McGraw-Hill Encyclopedia of Science & Technology*. Dr. Salmon is a fellow of the Society of Manufacturing Engineers.